The Engineer's Manual of Construction Site Planning

The Engineer's Manual of Construction Site Planning

Jüri Sutt

Professor of Construction Economics and Management
Tallinn University of Technology

Irene Lill

Professor and Head of Department of Building Production
Tallinn University of Technology

Olev Müürsepp

Associated Professor
Tallinn University of Technology

WILEY Blackwell

This edition first published 2013
© 2013 John Wiley & Sons, Ltd

Registered Office
John Wiley & Sons, Ltd, The Atrium, Southern Gate, Chichester, West Sussex, PO19 8SQ,
United Kingdom

Editorial Offices
9600 Garsington Road, Oxford, OX4 2DQ, United Kingdom.
The Atrium, Southern Gate, Chichester, West Sussex, PO19 8SQ, United Kingdom.

For details of our global editorial offices, for customer services and for information about how
to apply for permission to reuse the copyright material in this book please see our website at
www.wiley.com/wiley-blackwell.

Library of Congress Cataloging-in-Publication Data

Sutt, Jüri.
   The engineer's manual of construction site planning / Jüri Sutt, Irene Lill, Olev Müürsepp.
      pages   cm
   Includes index.
   ISBN 978-1-118-55609-2 (pbk.)
1. Building sites–Planning–Handbooks, manuals, etc.   2. Building–Superintendence–
Handbooks, manuals, etc.   3. Civil engineering–Handbooks, manuals, etc.   I. Lill, Irene.
II. Müürsepp, Olev, 1936–   III. Title.   IV. Title: Manual of construction site planning.
   TH375.S88 2013
   692'.1–dc23
                                            2013002862

A catalogue record for this book is available from the British Library.

Wiley also publishes its books in a variety of electronic formats. Some content that appears in
print may not be available in electronic books.

Cover image: © iStockphoto/urbanglimpses
Cover design by Meaden Creative

Set in 11/14pt Palatino by SPi Publisher Services, Pondicherry, India

1   2013

Contents

List of Figures   viii
List of Tables   x
About the Authors   xi
Preface   xiii

Introduction   1

Chapter 1: Initial data   5
1.1 The project (design) documentation   6
1.2 The bill of quantities and the bill of activities   7
1.3 Job descriptions and specifications   7
1.4 The contract conditions set out in the bidding
     invitation documents   8
1.5 The report of the construction site inspection   8

**Chapter 2: Outline of site management
planning in the bidding stage**   15
2.1 The goal   16
2.2 The explanatory note   16
2.3 Construction site layout   19
2.4 The construction time schedule   21
2.5 Cost estimation of temporary works
     and construction site set-up   23

**Chapter 3: Outline of site management
after contract signature**   28
3.1 The goal   29
3.2 Initial data   29
3.3 Construction site layout   30
3.4 Construction scheduling   35
3.5 Calculation of site work quantities and
     estimate of costs   46

Chapter 4: Suggestions for choosing construction cranes   51

4.1   General   52
4.2   Selection and positioning of tower cranes   53
4.3   Selection and impact areas of mobile cranes   77
4.4   Cranes working near overhead power lines   91
4.5   Hoist danger area   94
4.6   Operating cranes near buildings in use   95
4.7   Restrictions on crane work   97
4.8   Working in the danger area   98

Chapter 5: Suggestions for calculating resource requirements   99

5.1   Construction site temporary roads   100
5.2   Construction site storage   105
5.3   Temporary buildings   111
5.4   Temporary water supply   115
5.5   Temporary heating supply   116
5.6   Temporary power supply   121
5.7   Construction site lighting   126
5.8   Construction site transport   127
5.9   Load take up devices   130
5.10 Construction site fencing   135

Chapter 6: On-site safety requirements   137

6.1   General basics and responsibilities   138
6.2   The duties of building contractors   141
6.3   The obligations and rights of the labourer   144
6.4   Ensuring safety on the construction site   146

Chapter 7: Requirements for work equipment   155

7.1   General requirements   156
7.2   Mobile work equipment   158
7.3   Lifting devices   160
7.4   Dangers from energy   161
7.5   The usage of work equipment   163

7.6 Usage of work equipment for temporary
        work at height   164
7.7 Work with flammable and explosive materials   168

Chapter 8: Work healthcare   **169**

8.1 Allowable physical effort   170
8.2 The usage of personal protective equipment   170
8.3 Welfare facilities and first-aid   171

Appendix: Construction site layout symbols   173
Bibliography   177
Index   178

List of Figures

Figure 2.1	Site layout in the bidding stage	20
Figure 2.2	An example of a time schedule in the bidding stage	22
Figure 3.1	An example of construction site layout for the frame erection stage	34
Figure 3.2	Network model for construction	37
Figure 4.1	Drafting geometrical parameters for a tower crane	54
Figure 4.2	Tower crane Liebherr 550 EC-H40 Litronic radius and capacity chart	57
Figure 4.3	Cross-linking the tower crane to the axes of the building under construction	59
Figure 4.4	Positioning the crane track on the edge of an unsupported recess slope	60
Figure 4.5	Longitudinal linking of the tower crane with building under construction	63
Figure 4.6	Danger areas around the building	66
Figure 4.7	Boundaries of the danger area	66
Figure 4.8	The tower crane impact areas	69
Figure 4.9	Danger areas above the building	70
Figure 4.10	Simultaneous operation of two cranes on the same rail track	73
Figure 4.11	Simultaneous operation of two cranes positioned on opposite sides of the building	75
Figure 4.12	Simultaneous work of two cranes positioned between two buildings under construction	76
Figure 4.13	Calculating mobile crane minimum boom length	78
Figure 4.14	Assembling at an angle	81

Figure 4.15   Example of determining the assembly
parameters based on lifting capacity chart
for the RDK 25 crawler crane   85
Figure 4.16   Example of determining the assembly
parameters for the Liebherr LTM 1030
mobile crane   86
Figure 4.17   Positioning of mobile cranes at the edge of
unsupported recess slopes   88
Figure 4.18   The minimal acceptable horizontal
distance s_5 from the bottom edge of a recess
with an unsupported slope to the nearest
outrigger of the crane (m)   89
Figure 4.19   Danger area of mobile crane equipped
with boom fall prevention device   90
Figure 4.20   Surveillance and danger areas of aerial
power lines   91
Figure 4.21   Extent of the surveillance and danger area
of the electrical overhead power line   92
Figure 4.22   Safe positioning of mobile crane close
to overhead power lines   94
Figure 4.23   Conditions of operation for tower crane
near a building in service   96
Figure 5.1   Various kinds of construction site road   104
Figure 5.2   Double- and quadruple-branched slings   132

List of Tables

Table 2.1   Example form of construction site cost
            estimate during the bidding stage   26
Table 3.1   Example of construction work classification   44
Table 3.2   List of costs for temporary and building site
            management works   47
Table 4.1   Assembly parameters of precast elements
            and lifting parameters of tower crane   56
Table 4.2   Assembly parameters of precast elements   82
Table 4.3   Lifting parameters of chosen mobile cranes
            compared to the assembly parameters
            of precast elements   84
Table 5.1   Average space required for storage of
            construction materials   110
Table 5.2   Recommendations for surface lighting
            in construction   125

About the Authors

Jüri Sutt has nearly 50 years of experience in construction management as a practicing manager, researcher, consultant and lecturer which has included designing the construction technology for large mines in Siberia, a gas trunk pipeline in Libya and managing a construction firm. In 1965, he pioneered the use of IT in construction management research in Estonia. Between 1965 and 1980, J. Sutt was a member of several USSR scientific councils in the field of construction management, and from 1965 to 1978, he was the head of the Construction Management Department of Estonia's State Building Research Institute which developed scheduling and cost estimating IT systems that were widely used in the Soviet Union.

He has been an adviser to four ministers responsible for building during Estonia's transition to a free market economy and led working groups elaborating construction market regulations in the 1990s. In addition, he has provided consultancy services for clients' projects and contract management and has gained expertise in contract disputes in the last 15 years.

In 1960, J. Sutt qualified as a construction engineer. He was awarded the Candidate of Science degree in 1968 (equivalent to a PhD), and, in 1989, the Doctor of Science (habil.) in mathematical methods and IT in economics. The principal outcome of his research has been the methodology of IT simulating production – economic activities of construction firms enabling experimentation with different economic mechanisms and management strategies in construction enterprises.

Since 1989, he has been Professor of Construction Economics and Management at the Tallinn University of Technology.

Irene Lill graduated from Tallinn University of Technology as civil engineer, and defended her degrees in the same university (PhD and MSc in Economics). She has over 20 years of academic experience in the university. She has been working in research closely with Jüri Sutt, initially as professor and student and as good colleagues today. Since 2005, she has been professor and head of department of Building Production in Tallinn University of Technology.

Olev Müürsepp graduated from Tallinn University of Technology as a civil engineer. He has nearly ten years of experience working as a site and project manager in a construction enterprise and three years in a large design firm as a consulting engineer in the field of design of technology and organisation of construction. For 10 years, he has worked in the Construction Management Department of Estonia's State Building Research Institute as a researcher in the field of modelling technological and organisational decisions in civil engineering. In 1987, he defended his PhD in this specialist area of construction engineering. Since1991, he has worked as associated professor in Tallinn University of Technology.

Preface

This handbook deals with the problems of engineering preparation for building in a construction company, both during the bidding phase and after a contract has been concluded.

The handbook's recommendations can also be used in the design phase, when the building contractor is not yet selected. In this case, it has the aim of assuring the constructability of the designed building and of calculating a control estimate for the owner in order that bids can be weighted and contractors' potential duration of construction can be evaluated. In the design stage, the methods used are similar to those of the contractor in the bidding phase, when aggregated norms are used.

The key problems consist of identifying the composition of complex project organisation and level of detail of the initial data, the inspection of the construction site, compiling the construction site layout and the construction schedule, and the cost estimate of construction site expenses. Suggestions for calculating the resource allocation are presented: for the selection of cranes and lifting devices, the planning of temporary buildings and roads, and for technological networks, fire safety, fencing and lighting. On-site safety precautions in planning of the construction site management are discussed.

The owner's construction costs are determined through cooperation between the owner and the designer/consultant, according to preliminary design task as set out by the owner and the designer's technical and aesthetic competence. The structural designer must ensure the building's strength, stability, compliance with environmental criteria, etc. These costs are also affected by the detailed plan requirements validated by the local authorities. Another concern is that not enough attention is paid to construction management and building technology during the design of the construction contract conditions, and their subsequent negotiation. This, however, impacts the duration of construction, and based on this the contractor will be able to make the lowest price offer without reducing the quality of constructing. Often ignored is the fact that temporary works and temporary facilities on the building site form up to 12% of total costs, depending on the type of the building, site conditions, seasonality and the building owner's stipulations on duration.

This can be explained by the fact that construction site management and temporary facilities costs are not reflected in the final physical form of the building and will therefore remain unnoticed unless specially outlined by the consultant. Construction site management is reflected in the economic result of the owner's investment in the construction project, especially for business projects. The quicker the construction is completed, the sooner it becomes profitable.

For example, for a building that costs €100 million, with an annual profit rate of 10%, shortening the duration of construction would provide an additional monthly profit of approximately €0.8 million, and furthermore, it would enable the saving of about €0.5 million on the construction loan interest payments. Nevertheless, it should not be forgotten that for the contractor, this may entail organising the work into several shifts, bearing in mind winter conditions, etc., and the resulting additional costs will need to be compensated.

For this reason, the importance of the preparatory engineering work, called construction site management design, cannot be underestimated. Overall, it is divided into three phases:

❑ The project's main designer orders the construction site management project from a specialised consultancy company. The result forms the basis of the owner's financial plans (loan agreements) and the conditions of the contracts with designers and builders.

❑ The contractor prepares the construction site management project for calculation of bidding price and construction deadline.

❑ The firm that wins the competitive bidding process prepares the construction site management project consisting of the site plan and time schedule, at the same time calculating the cost price and compiling working drawings.

This handbook describes the specifics of the last two stages, bearing in mind that in the first stage, that is the design phase, the preparation of the construction site management project is similar to the contractors planning of site management in the bidding phase. However, it may be less detailed because the construction company is as yet unknown. However, how can the owner prepare a financial plan and predict the temporal parameters of the loan agreements without calculating the duration of construction? Preparing a time schedule requires a scheme plan of the site and temporary works. Preparing a construction site management project in the design phase certainly requires involvement of a specialised consultant or an impartial contractor.

This handbook is meant for planners of construction site management, construction engineers and construction site quantity surveyors, but also for students who specialise in civil engineering and construction.

The authors are grateful to J. R. Illingworth, D. J. Ferry, P. S. Brandon, H. Bauer, R. Salokangas, L. Dikman, F. Harris and R. McCaffer who have analysed different aspects of construction site management and inspired the authors of this handbook to approach the construction site problems from a different perspective – as a set of simultaneous problems.

In compiling the book, Jyri Orlov (MERKO AS), Taimo Kikkas and Enn Siim (Skanska EMV AS) helped the authors by providing useful hints and suggestions, and the authors are very thankful to them.

If there are discrepancies between recommendations given in the present handbook and prescriptions given in local laws, codes, instructions or standards, local regulations must be followed.

His co-authors - Irene Lill and Olev Müürsepp - and his publishers were saddened to hear of the death of Jüri Sutt, who passed away on April 20th 2013.

Introduction

The aim of construction site management planning is to find solutions to erect buildings in the cheapest, fastest and safest way possible, based on construction sketches and layouts, valid design and building standards, and on the owner's wishes concerning construction time and demands for the quality of the construction. Planning of site management is based on knowledge of building technology and different methods of the time scheduling of construction work.

To fulfil this goal, one must prepare:

- the budget of the construction expenses;

- the time schedule of construction works;

- the construction site layout(s);

- the cost estimate for the set-up of temporary buildings and site management;

- the list of risks.

In the methodological sense, this task entails the planning of alternative solutions from the viewpoints of building technology and site management, the assessment of those

The Engineer's Manual of Construction Site Planning, First Edition. Jüri Sutt, Irene Lill and Olev Müürsepp.
© 2013 John Wiley & Sons, Ltd. Published 2013 by John Wiley & Sons, Ltd.

solutions on the basis of the chosen criteria and, finally, selection between them.

In making the selection, the following evaluation criteria can be applied:

❑ the proportion (%) of the cost of the temporary buildings in relation to the general cost of the building complex, which in construction varies to a great extent (1.5–12%);

❑ duration of the construction period;

❑ the bill of quantities for temporary buildings, including their proportion within the overall cost of temporary works;

❑ the quantity (length, area) of temporary construction and their cost by type of construction (temporary roads, buildings, utility networks, etc.);

❑ the unit cost of temporary buildings and facilities per €1 million of construction cost, or per hectare of construction territory (used mainly during the construction pre-planning stage);

❑ total labour consumption of erecting temporary buildings in man-days (for construction preparation period separately), and unit quantity of work per unit area of construction, or per total cost of construction, or another parameter.

Distinguishing building technology and building management is by convention. By the planning of building technology we mean:

❑ the description of construction process in space (the plan and section of the construction site and/or work front);

❏ the description of the construction process and resource allocation in time (line charts or time-space charts);

❏ the work quality requirements;

❏ the allowed tolerances;

❏ the safety requirements, taking into the account working methods and tools.

By construction management we mean making separate works compatible with each other in order to erect a building as a whole, that is above all, the correlation between various construction works and processes, the conditions of preparing and handing over the job site, separate works and completed construction stages.

Keeping in mind the purposeful differences of each construction project at the development stage, we must separate the planning of building management into two different phases:

❏ bidding calculations; and

❏ after winning the bidding competition, preparation of a contract.

The solutions presented will be considerably more precisely detailed in the second phase because the actual field of production in a construction company is being dealt with – the planning of the more or less complex processes of building.

During the first phase of design, the issues and problems that have to be solved in the second phase should be identified.

This handbook deals with the methods of planning the building site management that are largely common in regular construction, above all in erecting buildings. It does not concern work management for special structures (line structures, water structures, power plant structures and chemical industry plants, etc.). Neither does it deal with the compilation of technological charts (instructions) for each individual building process, nor will it present a catalogue for technological charts.

The list of all the actions and the documents compiled as a result of the actions described in the guide is long, and this means that not all of these procedures may need to be performed or their results presented in the same thoroughness or formality in every project. Thus, the guide serves as a reminder, referring to issues where the construction company has to take a decision when it wants to take part in any particular project.

Chapter 1
Initial data

Chapter outline

1.1 The project (design) documentation

1.2 The bill of quantities and the bill of activities

1.3 Job descriptions and specifications

1.4 The contract conditions set out in the bidding invitation documents

1.5 The report of the construction site inspection

The Engineer's Manual of Construction Site Planning, First Edition. Jüri Sutt, Irene Lill and Olev Müürsepp.
© 2013 John Wiley & Sons, Ltd. Published 2013 by John Wiley & Sons, Ltd.

1.1 The project (design) documentation

For preparation of site management solutions and decision making, the following documents are necessary:

❑ the layout of the plot of land (the construction site situation plan), on which buildings under construction, existing buildings (including those due to be demolished) and utility networks, roads, paths, courts and geodetic data (including contours) are indicated;

❑ the plans and sections of buildings under construction;

❑ the head-note stating the general description of the project, the data of the architectural solution and the geological and hydrogeological conditions of the site;

❑ the list, location and capacity of existing utility networks, and those to be set up and demolished;

❑ the results of the project site survey, for example the availability and location of quarries, sources for supplying the construction site with electricity and water, the throughput of roads and bridges and various other documents.

The completeness of these data depends largely on the level to which the client/owner has resolved the tasks relating to the project survey and design phases of construction. In the call for tenders, it is advisable for the client to present the basic design, rather than only a building scheme design (brief), and other data in relatively limited format.

Here and later, we presume that the design of a construction investment project is divided into the following design stages:

❑ scheme design (brief), the basis of feasibility studies;

❑ preliminary design, the basis for permission to build from local authorities;

❑ basic design, the basis for construction works;

❑ working drawings – the engineering solution for complicated assemblies, which can include technological instructions.

1.2   The bill of quantities and the bill of activities

The bill of quantities should be an integral part of basic design and included in the bidding invitation documents (if the owner has ordered a bill in the contract to design). If a bill of quantities at the level of unit price is absent, then a bill by structural units and engineering facilities, with corresponding unit measures and physical capacities, must be used (part of the preliminary project). This list is called the bill of activities.

The contents of either the bill of activities or bill of quantities serve as the basis for assembly of the time schedule. If these bills are absent from the bidding invitation documents, they will be drafted by the construction company, ascertaining beforehand whether the client has any specific requirements for particular measurement instruction for the works, or for the classification of the construction costs presented in the bidding.

1.3   Job descriptions and specifications

Specifications are part of the bidding invitation documents, which need to be examined in order to determine their completeness; likewise, the client's particular requirements relating to building material, machinery or the quality of building works, which may necessitate special building technology, and equally the client's specific requirements concerning the storage or preparation of materials/products.

1.4   The contract conditions set out in the bidding invitation documents

Contract conditions might influence site expenses (deadline, duration of construction, design and building management, construction stages, restrictions on selection of subcontractors, etc.) and should be specified by the client in the bidding invitation documents.

1.5   The report of the construction site inspection

Before making the plan of the construction works and the calculations for bidding, one must become acquainted with the contract conditions, the project documentation, the bill of quantities and the bill of activities and specifications and undertake a site visit. The form of the land, its geological and hydrogeological conditions and the disposition of existing structures on the plot and in the vicinity might significantly influence the selection of building technology (including type, quantity and location of machinery on the site), the extent of construction costs (direct, as well as general, site-dependant costs), the duration of the construction and the probable risks. A representative of the client should also be present at the construction site inspection to answer any questions that may arise.

A report of the construction site inspection must be drafted, signed and dated. Photographs of the construction site will be added if necessary. Any questions in the report of the construction site inspection that require written answers should be included at this point. This handbook recommends using the following questionnaire. The bidder is free to add to the questionnaire depending on the project and on the conditions of the contract.

1) Access roads

- Are there any restrictions arising from the width, height or load-bearing capacity of access roads, bridges or overpasses?

- Could construction transport or machinery damage or litter the existing roads resulting in the need to pay compensation to the client, the local government or any third party?

- Is it necessary to access private premises in order to get to the construction site, and if so what would the costs be?

2) The conditions of construction site occupation

- Is it possible to use:

  o the existing roads or the underlay of designed roads as temporary roads?

  o existing sites to store materials and as set-aside ground reserves?

- What obstacles need to be dismantled (moved):

  o above ground (piping, wiring, trees, etc.)?

  o on the ground (piping, protected surfaces, etc.)?

- ○ underground (drainage, piping, cabling, old foundations)?

- ○ existing buildings and other structures?

- • What is the situation with regard to:

  - ○ trees (do they need preservation and protection, do they obstruct the work of construction machinery, is it necessary to measure their height)?

  - ○ objects (of antiquity, architecture, nature) under preservation and are there any resulting restrictions?

  - ○ bodies of water (is there a possibility of altering the water levels, or is there a need for bridging)?

- • What else needs to be done in the erection of temporary buildings and structures and construction site setup?

3) The boundaries of the construction site and adjacent areas

- • What kind of buildings and trees surround the construction site and the property? Measure their height to ensure they will not obstruct the working radius of the crane. Do they need protection, and if so, how?

- • Measure the distance of the building under construction from the construction site boundary or the existing buildings. Is it enough for the installation of lifting devices, movement of machinery and erection of scaffolding? Is it necessary to make any special arrangements (e.g. partial or complete closure of a road) in order to use the building technology planned?

- Ensuring the safety of outside staff or visitors:

  o Is it necessary to ensure passage on site for vehicles and/or people not associated with construction? Does this require special measures, for example construction of temporarily covered walkway in the danger area (crane, hoist and/or scaffolding)?

  o Is it necessary to build a temporary pavement and temporarily covered walkway on the fencing of the construction site?

  o Are there any kindergartens, schools, playgrounds in the adjacent area? What measures are necessary to ensure the safety of children? Does this require special measures for the construction work?

- Are there any bodies of water with altering water levels that might affect the construction works (e.g. needfor special dewatering measures in the area of construction excavation, strengthening of temporary roads, etc.)?

- Proximity of an airport to the construction site: might this restrict the height of the cranes, etc.?

- Presence of adjacent utility networks, for example electric and communication cables, piping: might they cause additional restrictions and risks?

- Problems arising from environmental protection: might they cause additional restrictions and risks?

  o The need to inform the public (in the vicinity); if yes, at what time?

4) Noise

The restrictions on the level of noise and its duration must be ascertained from local government. This is particularly important if there are schools, children's institutions or hospitals close by. There might be special restrictions to work during the evening and night.

5) Facilities for the supply of water and electricity for construction

Application for technical permissions for the supply of water and electricity for construction must be completed and submitted to the appropriate boards. Despite the allocation and connection of water and electricity for the erection of (permanent) buildings being agreed in the project documentation, the amount of water and electricity used during construction could be greater.

6) Soil, geological and hydrogeological conditions

Even if this data is stated in the building design documentation, the contract applicant should still inspect the construction site. When conducting an inspection during a dry period, one must not forget the possibility of change in conditions during heavy rain or in winter.

One can draw conclusions by observing the flora and also by questioning residents. Whether there is indication of soil contamination must also be ascertained.

7) Restrictions on working hours

When carrying out construction work in foreign countries, it is important to know the local restrictions on the length of the working week, the number of working hours per day and overtime hours. In addition, the dates of public holidays and possible collective vacations have to be determined.

8) Local weather conditions

Determining the weather conditions is vital in order to estimate possible time risks. Weather information is available from the local meteorological service and local residents.

9) Regulations set by the local authorities on building and recycling of materials

These activities involve:

- Determining detailed overall area plan and servitudes, which can influence the building site layout and / or construction time schedule;

- Competence-, technical-, financial- or other requirements to contractor according to law of local authorities;

- Peculiarities of registering the building according to local authorities;

- Regulations of using local raw construction materials;

- Local recycling regulations.

In case of building in foreign countries, it is compulsory before starting planning the building site to get familiar with the building law of the country; good construction practice; trade and crafts unions' regulations, etc., as these can influence the on site safety conditions and labour usage, marking the site, guard fencing and responsibility issues.

All the described activities of construction site inspection have a goal to minimise the cost of construction, its duration and the risk level as early as possible using methodology of engineering preparation of construction.

Chapter 2
Outline of site management planning in the bidding stage

Chapter outline

2.1   The goal

2.2   The explanatory note

2.3   Construction site layout

2.4   The construction time schedule

2.5   Cost estimation of temporary works and construction site set-up

The Engineer's Manual of Construction Site Planning, First Edition. Jüri Sutt, Irene Lill and Olev Müürsepp.
© 2013 John Wiley & Sons, Ltd. Published 2013 by John Wiley & Sons, Ltd.

2.1   The goal

The goal of construction site management planning in the bidding stage is to identify problems that may occur with construction from the point of view of the knowledge and resources of the construction company, and to estimate construction costs relating to the building site from the point of view of the requirements set out in the bidding invitation documents. The outline of building management in this stage does not represent a prescription of work, but rather the documentation necessary for bid preparation (cost and duration) or a reason for withdrawal.

The outline of site management in the bidding stage consists of the following documents:

❏ explanatory note;

❏ site layout sketch;

❏ general time schedule of construction works by neighbouring subcontractors;

❏ approximate estimate of site costs;

❏ list of site management issues requiring change or elaboration prior to conclusion of contract.

2.2   The explanatory note

The explanatory note briefly summarises the site management plans that will be presented to the bidding panel along with other documents mentioned earlier. The explanatory note contains:

1) the list of major buildings and facilities in the building complex;

2) a description of the relationship between the owner/client or the owner and the client, if they are not being represented as one person or institution;

3) the schemes of procurement (missing parts of basic design or working drawings) and price mechanism (fixed lump sum, fixed lump sum with added bill of quantities, target price with cost reimbursement, etc.);

4) recommendations on the selection of subcontractors;

5) the total costs of temporary works, the same as a percentage of total construction costs, and the deviation from the average compared to similar projects;

6) duration of construction, including:

   • duration desired by the owner;

   • rational duration concluded from the time schedule;

   • contractor's time in reserve if he thinks that work can be completed more quickly;

   • list and duration of actions to be performed in winter;

   • need for shift work (what kind of works, percentage of the total), the resulting increase in direct costs;

   • possibility, and rationale for, additional shortening of construction duration.

7) Problems related to:

- materials and products;

- labour;

- construction machinery;

- subcontractors.

8) Other risks, for example:

- inadequacy of geological explorations;

- uncertainty about the client's ability to pay;

- quality of the presented drawings of buildings and facilities, including their co-ordination;

- any contradictions between the drawings and the bill of quantities;

- instability of the electrical supply, possible antiquity surprises, etc.

9) Issues that might need adjustment after the contract has been signed. These could be:

- technical conditions and contracts of temporary water supply, sewerage and electrical supply;

- redesign of foundations and frameworks to identify any possible financial savings;

- search for a buyer for any spare soil or recyclable materials emerging during demolition, etc.

2.3   Construction site layout

In this stage of contract management, the construction site layout is drafted as a sketch. The basis for this can be the plot layout or the location plan of site structures, on which the objects necessary for decision making from the site management standpoint may be drawn in freehand:

❑ existing buildings and structures (buildings and utility networks) on the site, the need to relocate or demolish same during the site setup, their availability for use during construction works;

❑ crane movement areas and danger zones;

❑ access roads with remarks concerning their state of order or the location of any planned new access roads;

❑ temporary roads on the site;

❑ the storage locations of materials and structures;

❑ in the case of a narrowly confined construction site, storage possibilities outside the site should be laid down on the situation plan of the construction;

❑ temporary buildings (offices and rooms for workers);

❑ temporary facilities on the site. The possible conditions for connecting to the electrical network, and connecting water and sewerage to existing pipe-work;

❑ excavated soil and storage of set-aside earth on the construction site;

❑ possibilities for waste storage on the construction site;

❑ the fencing of the construction site.

Since the construction site layout is based on the general situation layout of the project, and the solutions presented are impossible to elaborate in detail in this stage, the layout is compiled at a scale of 1:1000, 1:2000 or 1:5000.

If required, a vertical section of the building should be added to the layout to evaluate crane measurements. An example of a construction site layout in the bidding stage is given in Figure 2.1.

When drafting the construction site layout, the following should be observed:

❑ coherency with other parts of the building outline (design documentation);

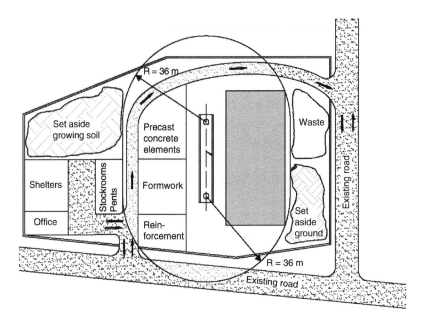

Figure 2.1:   Site layout in the bidding stage.

❑ accordance of construction works duration (the time schedule) with the chosen number of cranes and technological measures as planned on the site layout;

❑ the duration of construction pertinent to the time schedule (the number of cranes, etc. is dependent on this);

❑ the main building technology chosen;

❑ job safety requirements;

❑ fire safety requirements;

❑ environmental safety requirements;

❑ the goal for the lowest costs possible. This can be achieved by the help of:

  • the use of the buildings present on the construction site and those subject to demolition as temporary buildings while this does not interfere with the construction work,

  • the combining of temporary and permanent roads and sites,

  • the management of construction works according to as rational a scheme as possible, ruling out unreasonable accumulation of multiple works in a short time period, etc.

2.4   The construction time schedule

The following should be indicated separately on the construction time schedule:

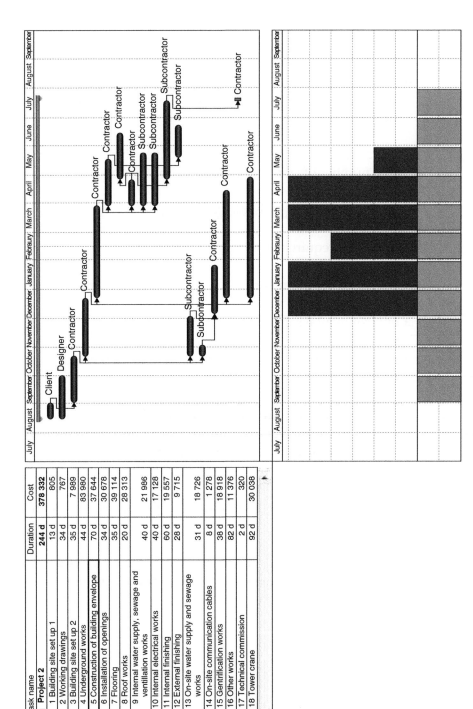

Task name	Duration	Cost
Project 2	**244 d**	**378 332**
1 Building site set up 1	13 d	805
2 Working drawings	34 d	767
3 Building site set up 2	35 d	7 989
4 Underground works	44 d	83 980
5 Construction of building envelope	70 d	37 644
6 Installation of openings	34 d	30 678
7 Flooring	35 d	39 114
8 Roof works	20 d	28 313
9 Internal water supply, sewage and ventillation works	40 d	21 986
10 Internal electrical works	40 d	17 128
11 Internal finishing	60 d	19 557
12 External finishing	28 d	9 715
13 On-site water supply and sewage works	31 d	18 726
14 On-site communication cables	8 d	1 278
15 Gentrification works	38 d	18 918
16 Other works	82 d	11 376
17 Technical commission	2 d	320
18 Tower crane	92 d	30 038

Figure 2.2: An example of a time schedule in the bidding stage.

❑ the works performed by the owner;

❑ the design works;

❑ the construction site set-up works;

❑ the building construction works listed by main structural elements, indicating separately the works performed by the contractor's own forces, the works that require erection and lifting machinery, and the works that require scaffolding,

❑ utility network construction works outside the construction site.

For every instance of work required, the duration in months (weeks), the number of workers and the number of shifts per day are given. An example of a time schedule used in the bidding stage is displayed in Figure 2.2.

2.5   Cost estimation of temporary works and construction site set-up

In this stage of cost estimation the following nomenclature of costs should be adopted. For every cost type, a corresponding normative unit measure is added, for example:

❑ costs for cranes and lifting machinery   €/day;

❑ costs for construction site fencing   €/m;

❑ costs for temporary roads and storage sites   €/m^2;

❑ costs for temporary water supply pipelines   €/m;

❏ costs for temporary sewerage facilities   €/m, €/day;

❏ costs for temporary electrical power
   distribution   €/m;

❏ costs for temporary buildings   €/m² × day;

❏ costs for construction site lighting   €/kW;

❏ costs for fire safety precautions   €/m²;

❏ costs for winter heating   €/m³ × day;

❏ costs for concrete maintenance in winter   €/m³ × day;

❏ costs for dewatering   €/day;

❏ costs for street and construction site upkeep   €/m²;

❏ costs for managing work on site   €/man-day;

❏ other costs   €/man-day.

When estimating costs, the company's own overall normatives are used with measurement units for each item. In small companies with no normatives, experiential appraisals are used.

The duration of the work, or of the use of service (crane work, heating of buildings, heating of concrete, dewatering), in days is gathered from the construction time schedule (Figure 2.2). Labour inputs in man-days are calculated on the basis of a graph of labour allocation according to the time schedule and the corresponding duration of work.

The areas (m²) requiring snow sweeping and street upkeep, the length of construction site fencing and utility networks (m) is measured on the site layout (Figure 2.1). The estimated heating cost for buildings is calculated from the cubage of the building and length of the cold period ($m^3 \times$ number of days); costs for clearing snow on the other hand are calculated from the area of temporary roads and sites and length of the winter ($m^2 \times$ number of days).

The cost norms (€ or physical units) for further use of necessary equipment, or the adjustment of existing norms, are calculated in the second stage of site management planning, that is on the basis of detailed cost estimates (or post-factum cost estimation) compiled after signing the contract.

This requires that construction site cost estimates are archived in the construction company and that the codes and units of cost item measurement are compiled consistently.

In addition to applying the construction site (temporary works) cost planning method in the bidding stage (as described), the construction company, whose works (buildings) and range are rather similar, can calculate construction site costs using even more widely aggregated norms.

For example, these norms could be the aforementioned 15-point summary divided either by:

❑ the cubage of the building   €/m³;

❑ the construction site area   €/m²;

❑ the duration of construction in days   €/day; or

❑ the cost of construction in direct expenses   €/€.

An example of construction site cost estimation during the bidding stage is presented in Table 2.1. This form also shows

Table 2.1:   Example form of construction site cost estimate during the bidding stage

....

(Name of the project)
Construction Site Cost Estimation
Cubage of buildings.. m³
Construction site area.. m²
Construction duration .... days
Cost of construction in direct expenses......................... €

Code	Type of cost	Measurement unit	Quantity	Price	Cost
1	Cranes and other lifting devices	Day/shift			
2	Construction site fencing	m			
3	Temporary roads and storage sites	m²			
4	Temporary water supply	m, m³			
5	Temporary sewerage	m			
6	Temporary power supply	m, kW			
7	Temporary buildings	m² × day			
8	Site lighting	kW, m²			
9	Fire safety	m³			
10	Heating the building in winter	m³ × day			
11	Concrete curing in winter	m³ × day			
12	Dewatering	days			
13	Streets and site upkeep	m² × day			
14	Managing costs on the site	man-day			
15	Other	€			
	Total	€			
	Costs for building volume	€/m³			
	Costs for construction site area	€/m²			
	Costs for duration of construction	€/day			
	Costs for construction direct cost	%			

the results of the aforementioned calculations of aggregated norms (the last four rows in the table).

The aggregated norms of productivity and cost price for temporary works units required to estimate the expenses described in this stage can be calculated on the basis of an analysis of the detailed estimates of such works on analogous past objects taken after winning the contract, when detailed unit price norms are used.

Chapter 3
Outline of site management after contract signature

Chapter outline

3.1 The goal

3.2 Initial data

3.3 Construction site layout

3.4 Construction scheduling

3.5 Calculation of site work quantities and estimate of costs

The Engineer's Manual of Construction Site Planning, First Edition. Jüri Sutt, Irene Lill
and Olev Müürsepp.
© 2013 John Wiley & Sons, Ltd. Published 2013 by John Wiley & Sons, Ltd.

3.1 The goal

The aim of site management planning after contract signature is to give a definite code of practice for the preparation of the construction site and the execution of construction work on the site. The aims are similar to those in the bidding stage, except that the construction site layout, the construction works time schedule and the estimate of costs are compiled in greater detail (see Table 2.1). The site management outline consists of the following documents:

❑ construction site layout(s);

❑ the technological model of construction, either as a network chart or, in the case of flow construction, a time-space chart (cyclogram);

❑ time schedule of the construction works;

❑ cost estimate of construction site costs;

❑ ordering calculations for resources in accordance with resource allocation for construction works;

❑ technological instructions for complicated construction processes (frame erection, piling, etc.).

3.2 Initial data

In addition to the site management outline drafted in the bidding stage, initial data includes:

❑ contract conditions;

❑ record of the bidding panel meeting;

❑ technical conditions and utility network integration contracts;

❑ list of subcontractors approved by the client (if contract conditions prescribe approval);

❑ decision on the technological scheme of the main construction process (critical path in network chart), including the number of working shifts;

❑ decision on the selection of a compiler of the working drawings.

3.3   Construction site layout

For large and complicated construction projects, site layouts are drafted for the different stages of construction; these could be:

❑ site setup stage;

❑ excavation works;

❑ foundation works;

❑ erection of frame;

❑ mounting heavy and complicated facilities both inside and outside the building;

❑ roofing works;

❑ finishing works;

❑ external civil engineering utilities networks;

❑ construction of water and sewage cleaning devices.

The layout is drafted at different stages because it is impossible to reflect schemes showing moving construction machinery and working teams, the placement of lifting equipment, sites for material storage, etc. from different phases of construction on a single ground plan.

On construction site layouts, the following elements are generally indicated:

❑ original surface contours and benchmarks on buildings and structures;

❑ drainage of rainfall (scheme);

❑ layout of temporary roads, their widths and structures;

❑ traffic map with necessary traffic signs;

❑ permanent and temporary buildings;

❑ permanent and temporary facilities (utility networks), depth and height of temporary utility networks, their connections with permanent networks and locations of wells, cells and switchgear;

❑ motion schemes of construction machinery used during erection of frame from precast elements;

❑ location of hoists;

❑ binding of tower crane to the axes of building and position at off-hour;

❑ vertical scheme of tower crane construction with the overall dimensions of the building (this can be presented as a separate drawing, see Figure 4.1);

- ❏ location of test load/check weight for the crane;

- ❏ space for storage of lifting devices and grapple equipment;

- ❏ location for receipt of concrete and mortar;

- ❏ storage places for set-aside excavated ground for backfill;

- ❏ warehouses and storage places for materials and precast elements by type;

- ❏ pre-assembly area;

- ❏ danger areas and their identification;

- ❏ locations of fire extinguishers and hydrants;

- ❏ location of switchboards (main switchboard, feed for the tower crane, etc.);

- ❏ location of power plant (transformer);

- ❏ protective fencing;

- ❏ location of lighting gantries and their heights;

- ❏ smoking area;

- ❏ waste containers/storage area;

- ❏ objects that require protection (high foliage, antiquities, etc.).

Possible sequence of procedures when drafting the construction site layout:

❑ choice of possible types of assembly cranes, their number and allocation, working and danger areas, places for storing materials and precast elements and also roads in the working range of cranes. These procedures are interrelated and must be solved at the same time;

❑ duration of construction in relative time is determined;

❑ resource allocation is determined on the basis of the time schedule (if the schedule fulfils the deadlines set in the contract), and includes:

- labour,

- output and voltage of the power supply (kW, V),

- water supply required (l/s),

- most important construction machinery, their main characteristics and quantity,

- precast elements and materials (average, maximum need per day, maximum need per shift);

❑ the need for temporary buildings, facilities and technological devices and their characteristics are calculated;

❑ construction site boundaries are indicated and a lighting scheme is drawn up.

Construction site layout is usually set out at a scale of 1:250 or 1:500.

An example of a construction site layout for the erection of a frame is given in Figure 3.1.

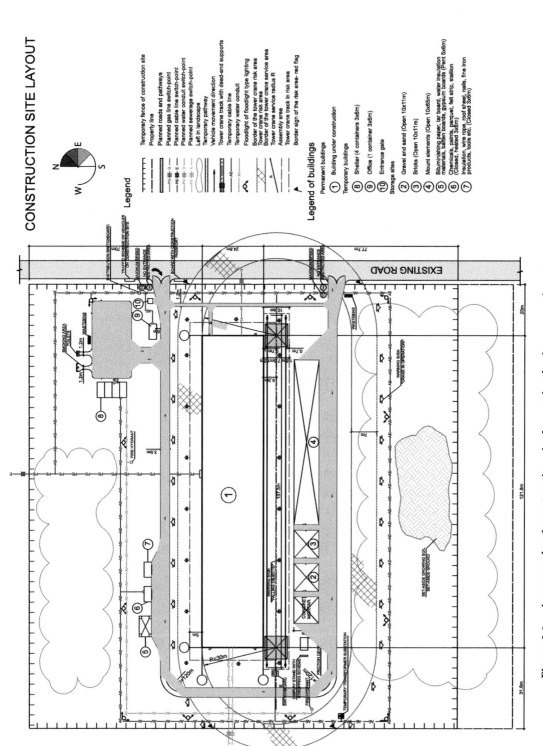

Figure 3.1: An example of construction site layout for the frame erection stage.

The symbols used in the construction site layout are in Appendix 1

3.4   Construction scheduling

Construction scheduling in this stage has the following purposes:

❑ a detailed description of requirements from the viewpoint of technology and construction management set at the time of bidding are taken as a basis, including:

- verification of the bill of quantities and scope of works,

- analysis and adjustment of planned building technology and site management,

- assessment of the intensity (number of workers engaged) and duration of works according to the adjusted scope,

- description of works sequence and relationships. In the case of larger and more complicated construction, advisably in the form of network or time-space chart;

❑ drafting of initial construction time schedule – using this schedule, the earliest starting and latest completion times of the subcontractor work presented in the call for tenders are determined;

❑ selection of building methods in order to secure the greatest retrenchment in the required construction duration.

This manual stresses that network or time-range charts are tools for the better elaboration of building technology and management, in which every item of work (or stage thereof) is associated with preceding or subsequent items. If a certain task can, technologically and from a management point of view, be initiated before completion of the directly preceding task, then the necessary readiness extent (in a chronologically calculated chart on which *moment of time* is event) of previous work must be identified in the chart.

The greatest advantage or freedom for the project manager in further construction time scheduling is available by this type of flow of work, with the beginning of all tasks in the network chart coming at the earliest possible time, and the possibility of completion at the latest, compared to other works. That is the best way to make use of slack time and to manipulate resources in the process of time scheduling. In other words, in this case the initial schedule of works (the non-chronological network or time-range chart) is flexible. In addition, the formation of the technological model of construction as a network chart will significantly save time for the construction manager later on, because during further planning, for example when drafting monthly schedules (during adjustment of the provisional overall plan), the initial chart does not generally require redrafting as the technological and organisational references initially set out in the network chart will ensure adherence to the essential sequences and references if the situation alters.

The second great advantage of the network and time-range models are the simplicity and speed of computing, which is important for formulating option solutions as well as adjusting the schedule of work during construction, and when considering the deviation of actual work (fill schedules) from the estimates.

The level of detail of the network chart should be chosen to represent:

❑ works and procedures of the owner that will be completed after signing the contract;

❑ the drafting of working drawings, if the construction begins prior to the end of the design work;

❑ construction site setup;

❑ all construction works for erection of buildings and structures, divided among the company's own working teams and all of the desired subcontractors;

❑ previously agreed benchmarks for interim financing.

An example of a network chart reflecting the relations and conditionality of construction works appears in Figure 3.2.

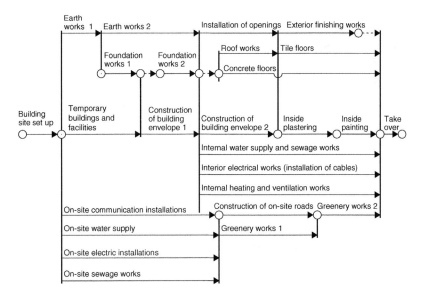

Figure 3.2   Network model for construction.

For every work included in a network model, the following data at least must be determined:

❑ name of each item of work (if work is divided into several parts/cycles, a number is added to the name);

❑ cost of work (from the budget);

❑ number of workers, which is determined by the project manager based on the size of the work front, and the conditions of the construction site and quantity of building machinery selected beforehand and their productivity, in order to assure a smooth servicing for workers;

❑ duration of work, determined on the basis of:

- quantities of works taken from the estimate (evaluated bill of quantities) and number of workers expertly selected to do their respective jobs,

- cost of work and the company's internal working efficiency (productivity) rate by works and the number of workers, or

- an expert appraisal;

❑ code by worker trade and/or subcontractors;

❑ number of working shifts per day;

❑ technological and organisational restrictions on work, for example seasonal demands and demand for separate works to be carried out simultaneously.

The compilation of the initial network model is followed by calculation of its chronological parameters in comparative time, that is the calculation is not linked to a calendar in order to:

❑ check the correlation between construction duration (duration of the critical path) of the created technological model and the time limit set by the contract agreement;

❑ determine the calculated early starting and late finishing times of works and the time float of works;

❑ get a better view of the technological and organisational relations of works, maximum extents of resource requirements (labour, materials, etc.) and their chronological divisions.

If the presented solutions satisfy the contract conditions, then the chart is linked to calendar dates giving the dates of the start of construction and hand over of the completed building as set out by the contract.

If the critical path turns out to be longer than the duration of construction set in the contract, then the technological and organisational solutions set as the basis for the network chart must be re-evaluated, verifying that the new solutions are feasible from the point of view of construction site conditions.

The principal ways to shorten construction duration are:

❑ selecting more efficient plant and machinery or to engage more workers;

❑ increasing the amount of machinery. This usually requires a large work front, which can be divided into working sections and allow various machines to be used simultaneously;

❏ division of the general work front into working sections, if the technological conditions and job safety allow it, to ensure earlier beginning of subsequent works;

❏ increasing the number of shifts.

As the shortening of the duration of construction is generally connected to a change in cost price, the following must be borne in mind:

❏ shortening the duration of construction is gained only by shortening the duration of tasks on the critical path;

❏ when the overall construction duration is gradually shortened, initially non-critical work chains on the network chart will become increasingly critical, that is the amount of tasks in need of shortening will grow equal to the square of the time shortened;

❏ when decreasing the duration of construction using a multi-shift division of work, the amortisation costs of construction machinery will decrease, while the costs for transport, assembly and disassembly of machinery will increase; however, running costs will stay constant (machine operator's wage, fuel consumption, costs for electricity, lubricants, repair and maintenance);

❏ working efficiency is up to 10% lower in the evening shift and up to 15% lower in the night shift, because:

  • there is inevitable loss of time when changing shifts,

  • inconvenience from evening and night work influences worker productivity (artificial lightning, etc.) and complicates coordination work with third parties,

- the number of on-the-job accidents increases,

- there might be a need for additional pay for working on evening and night shifts.

When deciding which option to choose in order to shorten construction duration, economic calculations should be followed taking into consideration the aforementioned, and other, substantial factors for each specific construction project.

After introducing the necessary changes to the work schedule, the calendared work schedule is calculated.

It follows from this that the compilation of the construction site layout and time schedule of construction works are mutually related, therefore finding a satisfying solution might take several iterations of the plan. At the same time, there must be a desire and will to be prepared for multiple calculations of resource allocation, the drafting of their workloads (workers, construction machinery, materials, etc.) and payment schedules and the cost estimates for construction site expenses.

As a result of the chronological planning of building management, the following documents are drafted:

- the organisational/technological model of construction (the network chart of construction);

- the construction work time schedule (Gantt chart);

- charts of labour allocation (as a histogram) separately, with works to be completed by the contractor's own forces, and in total;

- chart of basic plant allocation;

- chart of financing of works (cumulative) as an appendix to the contract in order to define contractual payment flow for the client.

On the basis of the chart of calendared construction work, the following should be indicated:

- deadlines of completion/hand over of missing drawings;

- duration of dewatering works;

- duration of the use of offices, shelters, warehouses and other temporary buildings;

- duration of the need for construction site fencing;

- duration of the need for safety barriers;

- duration of the use of scaffolding;

- working period of tower crane (and bigger mobile cranes) on construction site with reference to the need for support works as well as assembly and disassembly time;

- heating duration for construction of the building;

- duration of warming period for concrete;

- deadlines for delivery of technological equipment;

- connection dates for utility networks;

- testing dates;

- dates for inspections and expertise.

The construction company should create a classification system for works corresponding to its specialisation that elaborates on inter-company labour consumption norms or the reciprocal norms of labour productivity. The availability of this kind of data system would allow significant economies on costs and the time spent preparing construction time scheduling.

Table 3.1 presents an example of such a classification along with possible labour efficiency indicators. The numerical values should only be interpreted as an example illustrating the considerable variation of labour efficiency by type of work to assure the expediency of compiling such standards as well as the need for their periodic adjustment.

The technological/organisational solutions for erecting many buildings are similar due to their similar structural solutions. Therefore, it is practical for large construction companies and consulting site planning firms to form a catalogue of standardised network charts that represents the majority of technological and organisational descriptions of buildings. The catalogue of network charts of a custom main contractor usually consists of no more than ten to twenty types of charts.

An example of a list of types (they vary by composition of construction work, sequence and reference) is as follows:

❑ construction of a multi-storied framed apartment building;

❑ construction of a multi-storey cast-in-situ concrete apartment building;

❑ construction of multi-storey brick house;

❑ construction of a multi-storey office building;

Table 3.1: **Example of construction work classification**

Number or Code	Work	Labour productivity in € per man-day
1	Building site set up	80
2	Temporary buildings and facilities	90
3	Earth works (by machine)	238
4	Earth works (manual)	13
5	In situ concrete foundations	86
6	Precast concrete foundations	
7	Foundation for technological equipment	60
8	Backfilling (by machine)	142
9	Backfilling (manual)	14
10	Erection of precast concrete frame	
11	Erection of steel frame	159
12	Erection of precast sandwich panels	
13	Construction of building envelope	102
14	Installation of doors and gates	179
15	Installation of windows	110
16	Roof works	118
17	Construction of large concrete floors	
18	Construction of small concrete floors	52
19	Construction of tile floors	83
20	Construction of roll-material floors	123
21	Inside plastering works	17
22	Inside painting works	24
23	Inside tiling works	52
24	Exterior plastering works	23
25	Exterior painting works	21
26	Ventilation works	152
27	Internal water supply and sewage works	76

(*Continued*)

Table 3.1: (*Cont'd*)

Number or Code	Work	Labour productivity in € per man-day
28	Installation of sanitary ware	97
29	Heat insulation works	81
30	Interior electrical works/ installation of cables	93
31	Installation of lighting	169
32	Interior low-current works	86
33	Assembly of automatic equipment	144
34	Assembly of technical equipment	201
35	Site levelling	247
36	On-site water supply and sewage works	106
37	On-site heating pipelines	73
38	On-site electric installations	171
39	On-site communication installations	88
40	Construction of on-site roads and paved areas	139
41	Greenery works	113
42	Construction of fencing	88
43	Other works	76

❑ construction of a single-storey framed industrial building;

❑ construction of a multi-storey industrial building;

❑ construction of a petrol station;

❑ construction of a department store;

❑ construction of a waterworks;

❑ construction of a sewage treatment plant.

Construction time schedule for use in planning site management during the contracting period has to be more detailed than time schedule in bidding phase shown in Figure 2.1.

3.5 Calculation of site work quantities and estimate of costs

To simplify cost estimation of site planning and temporary works, the unification and standardisation of corresponding costs is necessary. For that purpose it is necessary to aggregate subsequent site work element costs estimated by unit prices (analogous to direct costs) into 15 standardised groups correlated with the nomenclature of the aggregated costs of the bidding stage (see Section 2.5). Therefore, feedback is used to automate the formation of norms for the first stage of site management.

This means that at this stage construction site costs are estimated as direct costs of the construction, given at the level of unit prices. The designer of the construction site management will compile a bill of quantities of temporary works and the respective list of necessary resources, which are estimated by the construction company's quantity surveyor.

The resource requirements appertaining to the bill of quantities should be given according to the resource ordering form used in the company, if such a form is created, and should be printed out separately.

Presented in Table 3.2 is a list of costs (resources) that refers to the initial data in order to determine costs according to the construction site layout (CSL), time schedule (TS) or to standards and procedures of calculating a bill of quantities. The abbreviations TS and CSL in the table point to the source from

Table 3.2:   List of costs for temporary and building site management works

Code		Cost Description of cost element and cost group	Measurement unit	Information source	
Group	Element			CSL	TS
1		*Cranes and other lifting devices*			
	1.1	Tower cranes (for calculations see Chapter 4)			
	1.1.1	Number and type of cranes	Machine shift	+	+
	1.1.2	Crane way base	m^2	+	
	1.1.3	Crane way track modules	Piece	+	
	1.1.4	Crane way safety fencing	m	+	
	1.1.5	Test load	t	+	
	1.1.6	Load take up devices by type	Piece	+	
	1.1.7	Switchboards	Piece	+	
	1.2	Mobile cranes	Machine shift	+	+
	1.3	Hoists and elevators	Machine shift	+	+
	1.4	Other lifting devices	Machine shift	+	+
2		*Construction site fencing*			
	2.1	Guard fencing	m; m^2	+	
	2.2	Safety fencing	m	+	
	2.3	Gates	Piece; m	+	
	2.4	Billboard with construction data (owner, contractor, designer, dates, etc.)	Piece; m^2	+	
	2.5	Protection barrier for trees	Piece; m^2	+	
3		*Temporary roads and storage sites*			
	3.1	Earthworks	m^2; m^3	+	
	3.2	Road covering	m^2; m^3		
	3.3	Open air storage covering	m^2; m^3		
	3.4	Location for receiving of concrete and mortar	Place		
	3.5	Pre-assembly area	m^2		
	3.6	Waste storage area, containers	m^2; piece	+	
	3.7	Traffic signs	piece	+	+

Table 3.2: *(Cont'd)*

Group	Element	Description of cost element and cost group	Measurement unit	CSL	TS
4		*Temporary water supply*			
	4.1	Common water pipes	m	+	
	4.2	Piping for fire brigade water	m	+	
	4.3	Fire brigade hydrants	piece	+	
	4.4	Water requirement for concrete maintenance	m³	+	+
	4.5	Water requirement for masonry works, plastering	m³	+	+
	4.6	Earthworks	m³	+	
5		*Temporary sewerage facilities*			
	5.1	Sewer ducting	m	+	
	5.2	Precipitation ducting	m	+	
	5.3	Earthworks	m³	+	
6		*Temporary power supply (for calculations see Section 5.6)*			
	6.1	Installable output	kW	+	
	6.2	Switchboards	Piece	+	
	6.3	Low voltage cable, wire	m	+	
	6.4	High voltage cable, wire	m	+	
	6.5	Aerial wire	m	+	
	6.6	Earthworks	m³	+	
	6.7	Supporting masts	piece	+	
	6.8	Transformers	Piece; kVA	+	
	6.9	Switchboards	Piece	+	
7		*Temporary buildings. People on construction site for maximum/average number of days (for calculations see Section 5.3)*			
	7.1	Office buildings	m²; day		+
	7.2	Dressing-rooms	m²; day		+
	7.3	Washrooms	m²; day		+
	7.4	Refectories	m²; day		+
	7.5	Drying room for clothes	m²; day		+
	7.6	Heated warehouses	m²; day		+
	7.7	Unheated warehouses	m²; day		+
	7.8	Open sheds	m²; day		+

(Continued)

Table 3.2: (*Cont'd*)

Code		Cost	Measurement unit	Information source	
Group	Element	Description of cost element and cost group		CSL	TS
	7.9	Toilets	Number; day		+
	7.10	Showers	Number; day		+
	7.11	Women's sanitary rooms	m²		+
	7.12	Smoking rooms	m²		+
8		*Construction site lighting (calculations see Section 5.7)*			
	8.1	Surveillance lighting	kWh; lux	+	+
	8.2	Road and site lighting	kWh; lux	+	+
	8.3	Working heading lighting	kWh; lux	+	+
	8.4	Lights and floodlights	Piece	+	+
	8.5	Light and floodlight posts	Piece	+	+
	8.6	Lighting cabling	m	+	+
9		*Fire safety*			
	9.1	Foam and powder extinguishers	Sets	+	
	9.2	Fire extinguisher equipment	Sets	+	
10		*Heating the building in winter*			
	10.1	Heating energy (m³ of room, heated days)	KWh	+	+
	10.2	Heating equipment	Piece; days	+	+
11		*Concrete maintenance*			
	11.1	Concrete heating	kWh		+
	11.2	Chemical admixtures	kg		+
	11.3	Concrete moistening in summer by sprinkling	m³ water		+
12		*Dewatering*			
	12.1	Pumps	Piece; days	+	+
	12.2	Electricity	kWh		+
	12.3	Water pipes	m	+	
13		*Streets and site upkeep*			
	13.1	Construction site cleaning	m²; weeks	+	+
	13.2	Street upkeep	m²; days	+	+
	13.3	Clearing of snow in winter	m²; days	+	+
14		*Managing costs on construction site*			
	14.1	Job position	man-day		+
15		*Other*			

which the amount can be found or measured. It is an illustrative list of costs of temporary works or corresponding vital resources in general construction, thus is not necessarily complete. The first column, numeration (code), corresponds to the cost group codes relating to temporary works estimates in the bidding stage (see Table 2.1).

The nomenclature of costs and resources presented in Chapters 2 and 3 can be regarded as an example of the typical temporary works that usually occur in dwelling and office building projects, as shown in Table 3.2. It is expedient for every building company or potential group to elaborate on the analogous two-level classification for themselves, taking their own specialisations into account.

Chapter 4
Suggestions for choosing construction cranes

Chapter outline

4.1 General

4.2 Selection and positioning of tower cranes

    4.2.1 Selection of tower cranes

    4.2.2 Positioning the crane

    4.2.3 Crane impact areas

    4.2.4 Using several tower cranes simultaneously

4.3 Selection and impact areas of mobile cranes

    4.3.1 Selection of mobile crane

    4.3.2 Work of the mobile crane by a recess

    4.3.3 Mobile crane impact range

The Engineer's Manual of Construction Site Planning, First Edition. Jüri Sutt, Irene Lill and Olev Müürsepp.
© 2013 John Wiley & Sons, Ltd. Published 2013 by John Wiley & Sons, Ltd.

4.4   Cranes working near overhead power lines

4.5   Hoist danger area

4.6   Operating cranes near buildings in use

4.7   Restrictions on crane work

4.8   Working in the danger area

4.1   General

The goal of this chapter is to describe the principles of positioning construction cranes in the safest possible way. The chapter explains the restrictions, distances and measures one should consider while planning the construction site. The rules and calculations described in the following text guarantee the safe coordination of cranes and personnel on the site. These suggestions should be taken into consideration where construction site conditions allow. Unfortunately this is not applicable in all cases and it is certainly possible for several cranes to work simultaneously in close proximity.

All complicated situations, where following these safety requirements is impossible, have to be approached case by case. In these circumstances, special instructions for cranes and personnel have to be compiled and supervision provided for all parties involved. It is important to understand that these instructions should not be general narratives but carefully calculated guidelines with danger distances, their identification signs, behaviour routines, etc. Every employee involved should understand the essence of different danger zones and what exactly they should do or be aware of before raising the hook.

4.2   Selection and positioning of tower cranes

4.2.1   Selection of tower cranes

First the required lifting parameters of the crane (lifting capacity, lifting height and radius) are determined, followed by the position of the crane and its working and danger areas with reference to the construction site conditions and possible restrictions. The distance between the building under construction and existing buildings, as well as safety requirements might affect the position and selection of the type of crane.

The lifting height and radius are determined by the chart in Figure 4.1.

The overall dimension of the building (and the parameters of the slope of the foundation recess, if necessary) and the assembly parameters of precast elements are taken as a starting point. The vertical chart should be presented on the construction site layout (CSL) or on a separate sheet.

The assembling height, that is the maximum required height of the hook H_{max}, is calculated as follows:

$$H_{max} = h_1 + h_2 + h_3 + h_4 \qquad (4.1)$$

where
h_1 – the mounting height of the assembled unit measured from the standing level of the crane, in m;
h_2 – over lifting height (usually taken as 0.5 m);
h_3 – the height of the assembled unit, in m;
h_4 – load take up device height, in m.

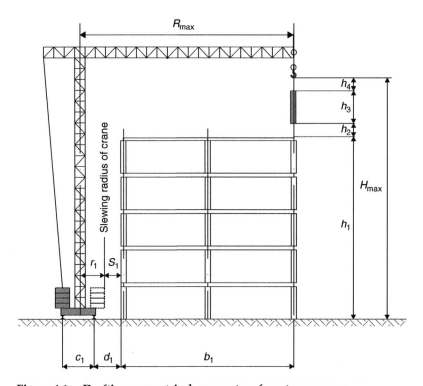

Figure 4.1:   Drafting geometrical parameters for a tower crane.

The required radius R_{max} of the crane depends on the farthest assembled element and possibilities of positioning the tower crane, as follows:

$$R_{max} = \frac{c_1}{2} + d_1 + b_1 \qquad (4.2)$$

where
c_1 – distance between the rails of the crane, in m;
d_1 – distance between the closest part of the building and the nearest rail, in m;
b_1 – distance between the farthest assembled unit and the closest part of the building towards the crane, in m.

The required lifting capacity (in tons) is determined for the placement of various heavy precast elements in the most difficult lifting conditions of the crane. For this purpose, the heaviest and furthest elements from the standing position are chosen, and their assembly parameters calculated. These results should be presented in the form presented in Table 4.1.

The assembly weight G_{max} is calculated as follows:

$$G_{max} = g_1 + g_2 \qquad (4.3)$$

where
g_1 – the weight of lifted precast elements with necessary devices (i.e. pre-mounted working platforms, supports, etc.) if applicable, in t;
g_2 – weight of load take up device with mass of hook and hook traverse, in t.

The tower crane is chosen on the basis of a comparison between the assembly parameters of the elements hoisted and the lifting parameters of the crane, as shown in Table 4.1. An example of presenting technical data of a suitable tower crane for particular precast elements is presented in Figure 4.2.

Here it should be borne in mind that determining the type of crane and its lifting capacity and the geometrical linking of the crane track to the building axes are iterative processes.

4.2.2 Positioning the crane

Two problems must be solved when cross-linking the tower crane to the building axes:

❑ determining the minimum allowable distance between the crane track axis and the closest longitudinal axis of the building;

Table 4.1 Assembly parameters of precast elements and lifting parameters of tower crane

No.	Precast concrete element	Assembly parameters of precast elements									Lifting parameters of the crane						
		Assembly weight (t)			Assembly height (m)					Assembly radius (m)	Trademark and technical data	Selected working parameters					
		Element $g1$	Load take up device g_2	Total G_{max}	Mounting height h_1	Over lifting height h_2	Element h_3	Load take up device h_4	Total H_{max}	R_{max}		Tower height (m)	Maximum radius (m)	Working radius (m)	Lifting capacity (t)	Lifting height (m)	
1	2	3	4	5	6	7	8	9	10	11	12	13	14	15	16	17	
1	Wall panel SW-110	13.0	0.4	13.4	34.0	0.5	7.8	2.5	44.8	35	Tower crane Liebherr 550 EC-H40 Litronic	8.5 m + 7 sections × 8.5 m	40	35	18.29	49.9	
2	Wall panel SW-213	10	0.4	10.4	23.4	0.5	4.3	1.5	29.7	40	Working radius • max 40 m • min 4.3 m			40	15.60		
3	Beam	20.9	2.4	23.3	25.6	0.5	0.4	6.5	33.0	25	Lifting capacity • with max radius 15.6 t • with min radius 40 t			25	26.74		
4	TT-slab	13.9	1.3	15.2	16.3	0.5	0.3	6.0	23.1	30	Hoisting height: max 49.9 m			30	21.84		

Notes
1. When determining the mounting height of the element (see column 6), the actual standing level of the crane measured from its bearing surface must be considered, not solely the assembly height set in the project.
2. The assembly radius of the crane (column 11) depends on the assembly weight and the chosen assembly scheme – if one or several elements are mounted from one position, etc.

❑ determining the range of the crane service area towards the lateral axis of the building from the viewpoint of positioning the crane in relation to the building.

Tower cranes moving on rails must be positioned next to the building under construction in order to comply with safety requirements, that is there must be safe distance between the closest parts of the building and the crane, and the edge of the

(a)

Hubhöhe Hoisting height / Hauteur sous crochet / Altezza di sollevamento
Altura bajo gancho / Altura de montagem / Высота подъема

C 25				500 HC
12	70.8*	–		–
11	65.0	73.1*		77.2*
10	59.2	67.3*		71.4
9	53.4	61.5		65.6
8	47.6	55.7		59.8
7	41.8	49.9		54.0
6	36.0	44.1		48.2
5	30.2	38.3		42.4
4	24.4	32.5		36.6
3	18.6	26.7		30.8
2	12.8	20.9		25.0
1	7.0	15.1		19.2
0	–	9.3		13.4
	m			m

Figure 4.2:   Tower crane Liebherr 550 EC-H40 Litronic radius and capacity chart: (a) detecting tower height; (b) detecting lifting capacity. *Further hoist heights and jib lengths as well as climbing inside the building can be obtained on request.

Ausladung und Tragfähigkeit
Radius and capacity / Portée et charge / Sbraccio e portata / Alcances y cargas / Alcance e capacidade de carga / Вылет и грузоподъемность

(b)

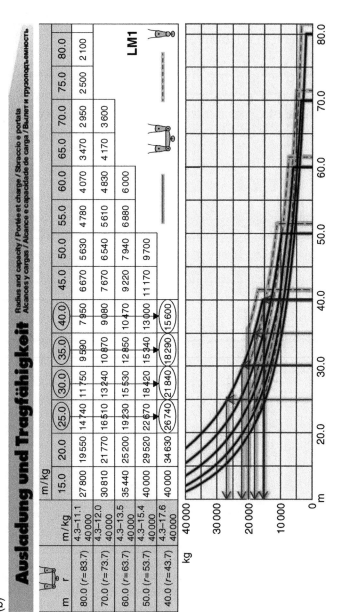

m r	m/kg	15.0	20.0	(25.0)	(30.0)	(35.0)	(40.0)	45.0	50.0	55.0	60.0	65.0	70.0	75.0	80.0
80.0 (r=83.7)	4.3–11.1 40 000	27 800	19 550	14 740	11 750	9 590	7 950	6 670	5 630	4 780	4 070	3 470	2 950	2 500	2 100
70.0 (r=73.7)	4.3–12.0 40 000	30 810	21 770	16 510	13 240	10 870	9 080	7 670	6 540	5 610	4 830	4 170	3 600		
60.0 (r=63.7)	4.3–13.5 40 000	35 440	25 200	19 230	15 530	12 850	10 470	9 220	7 940	6 880	6 000				
50.0 (r=53.7)	4.3–15.4 40 000	40 000	29 520	22 670	18 420	15 340	13 000	11 170	9 700						
40.0 (r=43.7)	4.3–17.6 40 000	40 000	34 630	(26 740)	(21 840)	(18 290)	(15 600)								

LM1

Figure 4.2: (Cont'd).

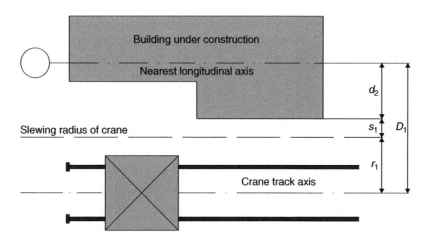

Figure 4.3: Cross-linking the tower crane to the axes of the building under construction.

crane way underlay must be outside the collapsing prism of the recess slope.

The distance D_1 of the crane track axis from the nearest longitudinal axis of the building is presented in Figure 4.3 and is calculated as follows:

$$D_1 = r_1 + s_1 + d_2 \qquad (4.4)$$

where
r_1 – slewing radius of the crane base (or other farthest part) (according to rating plate of the crane), in m;
s_1 – safety distance between the outside of the building, pile of precast elements, etc., and the farthest part of the crane.

❑ at up to 2 m height from the ground $s_1 \geq 0.7$ m;

❑ at heights over 2 m, $s_1 \geq 0.4$ m;

d_2 – distance from the closest part of the building (the outside wall) to the building's longitudinal axis nearest to the crane, in m.

Note. The crane track can be built only on the grounds of a ratified outline (project solution).

When setting the crane track in the proximity of a recess or trench with unsupported sides, their depth h_5 must be taken into account along with the soil grain so that the edge of the crane track underlay nearest to the recess would be outside the collapsing prism of the recess slope, as shown in Figure 4.4, where:

c_1 – distance between crane rails, in m;
h_5 – depth of recess, or trench width, in m;
h_6 – thickness of underlay (ballast) under the crane track (dependant on material and crane used), in m;
c_2 – width of crane way embankment, in m;
d_3 – minimum horizontal distance between the lower edge of the recess slope and the lower edge of the crane track underlay (ballast), in m;

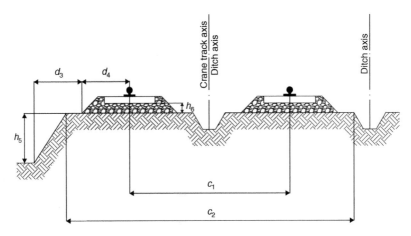

Figure 4.4: Positioning the crane track on the edge of an unsupported recess slope.

d_4 – distance between the lower edge of the crane track underlay and rail axis, in m;

The minimum distance d_3 along the horizontal between the lower edge of the recess slope and the lower edge of the crane track underlay (ballast) depends on the depth of recess, the soil and stability angle of the recess slope and can be taken approximately $d_3 \geq (1.0 \dots 1.5) \, h_5 + 0.4$ m.

While determining the distance d_4 between the lower edge of the crane track underlay and rail axis, the particular parameters and requirements of chosen crane should be considered.

When longitudinally linking the crane to the building under construction, the following must be determined:

1) the outermost stopping points of the crane in relation to the ends of the building;

2) the necessary length of the crane track.

Prior to the longitudinal link, the cross-linking of the crane must be completed, that is the location of the axis of the crane track has to be determined and executed as per the CSL.

The outermost stopping points of the crane are calculated as follows:

❑ On the opposite side of the building from the point of the crane, notes are drawn on the axis of the crane track from the building's outermost corners to a distance equal to the maximum radius of the crane;

❑ From the middle of the side of the building nearest to the crane, two marks are drawn on the axis of the crane track at a distance equal to the minimum reach of the lifting hook;

❑ From the centre of gravity of the heaviest units (their design position), marks are drawn on the axis of the crane track at a distance determined by the greatest load moment of the crane.

The outermost marks made on the axis of the crane track will determine the outermost stopping points of the crane. On the basis of the outermost stopping points of the crane, it is possible to calculate the length of the crane track L_1 as follows:

$$L_1 = l_1 + \frac{c_{cr}}{2} + 2(c_4 + c_5 + c_6) \qquad (4.5)$$

or approximately

$$L_1 = l_1 + c_{cr} + 4\,\text{m} \qquad (4.6)$$

where
l_1 – distance between the outermost stopping points of the crane, in m;
c_{cr} – width of the crane undercarriage, found in reference books, in m;
c_4 – distance between the end of the rail and the cul-de-sac (driving limiter), $c_4 = 0.5\,\text{m}$;
c_5 – breaking distance of the crane, at least 1.5 m;
c_6 – distance between the bumper and outer edge of the undercarriage from reference books, in m.

The calculated length of the crane track is adjusted upwards depending on the length of the track way link according to producer.

The longitudinal linking of the tower crane to the building under construction is presented in Figure 4.5.

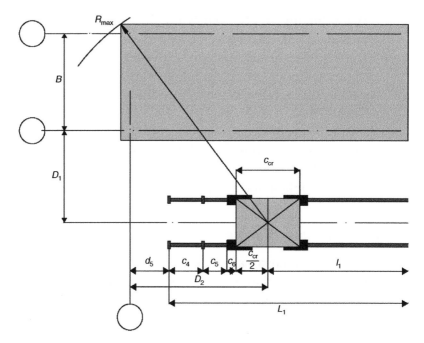

Figure 4.5: Longitudinal linking of the tower crane with building under construction.

The distance between the outermost working position of the crane and the last lateral axis of the building can be calculated as follows:

$$D_2 = d_5 + c_4 + c_5 + c_6 + \frac{c_{cr}}{2} \qquad (4.7)$$

where

d_5 – distance between the end of rail and last lateral axis of the building, in m;

$c_{cr}/2$ – distance between the outer edge of the undercarriage and the crane axis, in m;

B – distance from the outer longitudinal axes of the building and the closest to the crane axes, in m.

When linking safety guards for the crane track, it is vital to provide a safe distance between the crane elements and the fencing. The safety distance between the slewing radius of the crane undercarriage or other overhanging part of the crane (taken from reference books) and the safety fence should be at least 0.7 m.

The outermost stopping points of the crane should be drawn on the CSL and marked on the ground so that the markings are clearly visible to the crane operator and slinger.

4.2.3　Crane impact areas

In three-dimensional planning of the construction site and particularly in planning the positioning of construction machinery, the risk areas for people must be determined, where danger factors are present either temporarily or continually.

Continuous danger factors occur where the displacement of loads takes place with the help of lifting devices (assembling and loading machinery). Such areas must be surrounded by safety or signal fences. The meaning of safety fences here is structures that prevent an outsider accidentally gaining access to the dangerous area.

The impact range of danger factors is around the building and its floors, and within the working area of the crane, where assembly and demolition of building components takes place. These areas are surrounded with signal fencing. The meaning of signal fencing here is structures that caution against danger factors and mark the areas of restricted access on the construction site.

When working in these areas, special organisational and technical precautions must be applied that ensure safety.

Various areas are distinguished from the job safety point of view, including:

❑ assembly area of the building;

❑ working or service area of the crane;

❑ load movement area;

❑ crane danger area;

❑ danger area of roads (including crane tracks);

❑ danger area above the building.

The danger areas around the building are presented in Figure 4.6.

Assembly area here means the land surrounding the building, wherein assembled elements or units could fall. This area should be marked on the CSL. The assembly area should be considered potentially dangerous. For a building up to 20 m high, the width of the area s is 5 m. If the building is higher, the width increases as shown in Figure 4.7. Materials must not be stored in the assembly area, or in the crane track area isolated by signal fencing.

For an operating crane, the assembly area of the building is also part of the crane's danger area. The boundary of the assembly area is marked on the CSL, for example as shown in Figure 4.6, and with clearly visible warning signs on the construction site. Only assembly cranes and lifting machinery can be placed within these boundaries.

Certain places have to be allocated for people to enter, advisably on the opposite side of the building in relation to the tower

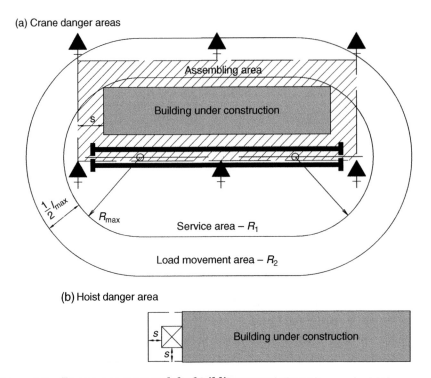

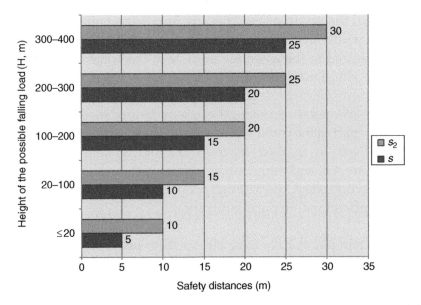

Figure 4.6: **Danger areas around the building.**

Figure 4.7: **Boundaries of the danger area** s – distance from the outer contour of the **building under construction;** s_2 – distance from the horizontal projection of the largest **overall dimension of the lifted load.**

crane, and these will be marked out on the CSL. In the building site area, passages within the assembly area must be covered with pents (see Figure 3.1).

The service area (working area) of the crane R_1 refers to the land that is within the boundary drawn by the crane hook when moving an assembled unit. In the case of a tower crane, this will be determined on the CSL by semicircles equal to the maximum reach of the jib R_{max} necessary for assembly in the outermost working positions of the crane, and the connecting straight lines in case there are no limitations on the moving range of the load, which might derive from construction site conditions.

The load movement area R_2 refers to the area where the farthest end of an assembled unit of maximum length hanging from the crane hook can move. The width of the load movement area of the tower crane equals the sum of the maximum reach R_{max} of the crane hook plus half the length of the longest lifted element l_{max}, presuming that the working range of the crane is unrestricted:

$$R_2 = R_{max} + \frac{1}{2}l_{max} \qquad (4.8)$$

where
R_{max} – the maximum reach of the jib when the crane is working, in m;

$\frac{1}{2}l_{max}$ – half the length of the lifted element with the largest overall dimensions, in m;

The load movement area is usually not indicated on the CSL, rather it constitutes only part of the danger area of the crane.

The risk area of the crane refers to the area within which the removable load (part) may fall to the ground, taking into consideration possible deviation (dispersion) from the vertical when falling.

The width of the tower crane danger area is determined using the following equation:

$$R_3 = R_{max} + \frac{1}{2}l_{max} + s_2 \tag{4.9}$$

where
s_2 – the width of the additional danger area deriving from the height of the assembly works according to construction regulations (see indicative values in Figure 4.7). This term reflects the possible deviation from the vertical (dispersion) when falling and depends on the lifting height and the dynamics of the load's motion (crane hook motion, squalls, etc.).

The hoist danger area s also depends on the height of construction and is presented in Figure 4.6b.

Impact areas of the tower crane in vertical section are presented in Figure 4.8.

The danger area over the building during construction of its upper floors is characterised by Figure 4.9.

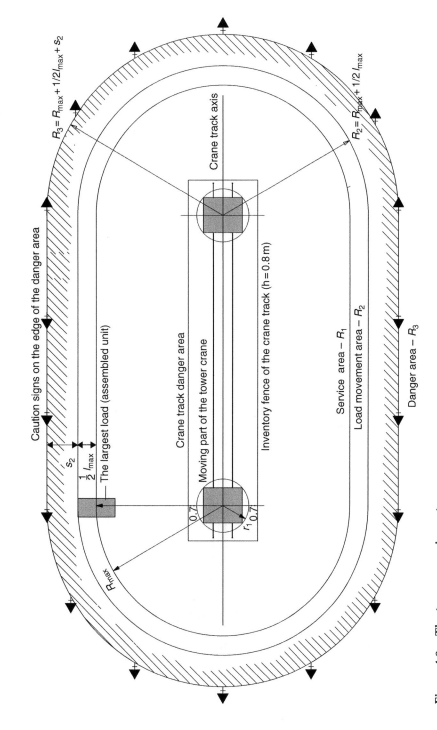

Figure 4.8: The tower crane impact areas.

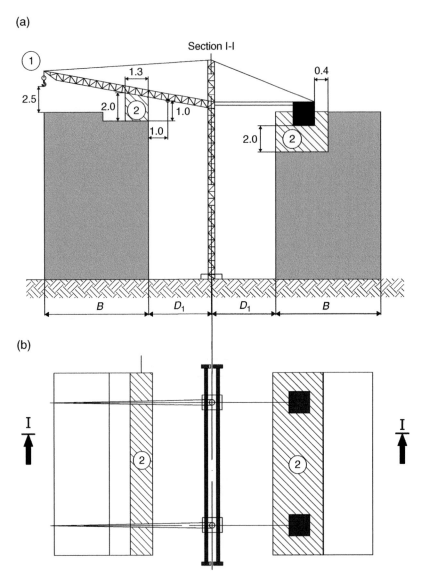

Figure 4.9: Danger areas above the building (a) for maximum reach of the jib; (b) for counterweight motion over assembly area; 1, position of the jib in cases of maximum jib reach; 2, danger zone (hatched areas).

During construction of the upper floors, the following safety recommendations should be considered:

❏ the space between the lifting hook and the assembly level should not be less than 2.5 m (Figure 4.9a)

❏ the space between the crane hook and the building element nearest to it must be at least 1 m across the horizontal and vertical (Figure 4.9a)

❏ the space between the lowest point of the cradle of the crane's counter weight and the assembly level must be at least 2 m (Figure 4.9b)

❏ The space between the farthest point of the crane's counter weight and the outermost protrusion of the building cannot be less than 0.4 m across the horizontal at a height of over 2 m from assembly or ground level (Figure 4.9b).

The danger areas that develop over the building are drawn on the CSL during the vertical linking of the crane, but similarly they are drawn on the technological map, if such is compiled.

4.2.4 Using several tower cranes simultaneously

When drafting the plan for construction works, there are different possible options for tying the assembly cranes to the building under construction. These options vary according to crane type as well as to the number of cranes used simultaneously.

The basis for selecting the number of cranes is generally:

❏ the estimated spatial parameters of the building under construction: its width, length and height;

❑ the quantities of assembly works; and the commissioning deadline of the building, that is the duration of construction.

The location and quantities of other important construction site elements, such as temporary roads and storage sites, etc., are dependent on the type and quantity of cranes on the construction site.

For long rectangular-shaped buildings, the tower cranes are positioned either on one or two sides of the building, depending on the width of the building, the lifting parameters of the cranes and the construction site conditions.

Two or more cranes, positioned on opposite sides of the building, are used when the reach of a crane jib, or the crane's lifting capacity, does not allow placement of all precast elements on one side of the building, or when one crane cannot guarantee the assembly capacity necessary to complete the building on time.

To avoid a collision between tower cranes moving on the same crane track, limit switches must be installed to the cranes' undercarriages. These switches must stop the cranes when the distance between the ends of the removable units with greatest length is <5 m.

For long buildings the building under construction and the crane track are divided into several zones (cycles). The length of each zone should not be less than double the working radius of the crane plus 5 m. Within the limits of each zone only one crane is generally allowed to work, the other crane must work in another zone or stand still with the boom turned in the opposite direction.

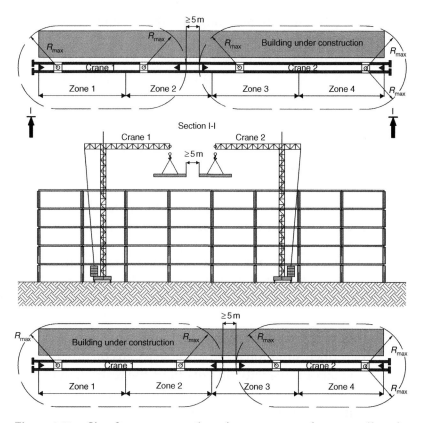

Figure 4.10:   Simultaneous operation of two cranes on the same rail track.

In Figure 4.10, the crane track is divided into four zones. In this case, when operating with two cranes the following must be taken into account:

❑ If crane no. 1 works in the first zone, then crane no. 2 can work in the third and fourth zones.

❑ If crane no. 1 works in the first and second zones, then crane no. 2 can work only in the fourth zone.

The advantage of mounting two tower cranes on one track is that:

❑ the overall length of the crane tracks decreases, compared to the cranes operating on two sides of the building;

❑ there would be no need for storage spaces and access to them on the other side of the building;

❑ the cost for electric supply to cranes decreases.

It is possible to place hoists for lifting materials as well as people on the other side of the building during construction.

A deficiency of this crane positioning is that it is a relatively more complicated arrangement of the simultaneous work of two cranes with reference to provide job safety.

The requirements for the positioning and safe working of the described situation are applicable when construction deadlines and labour intensity do not oblige the restriction of the crane's operating area; otherwise, detailed mounting instructions and schemes must be worked out for various time periods (for one or several shifts) and timely and proper notification provided to crane operators and workers.

Another option for two cranes to work simultaneously is presented in Figure 4.11. This option is used when:

❑ the width of the building exceeds the crane's jib outreach; or

❑ the load moment necessary to mount an element is greater than the allowed load moment of the crane.

The load moment necessary to assemble a unit is determined by multiplying the distance between the centre of the unit's

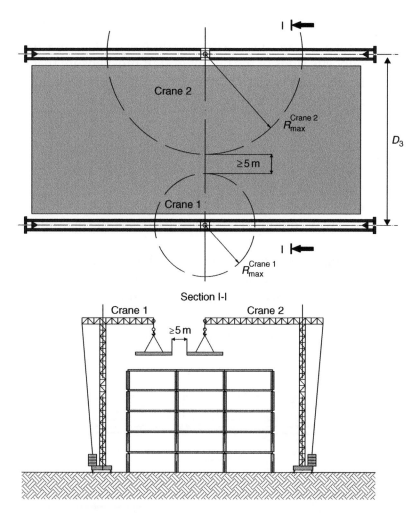

Figure 4.11: **Simultaneous operation of two cranes positioned on opposite sides of the building.**

weight and the crane axis, and the assembly weight (including the load take up device).

The crane positioning scheme and safety distances presented in Figure 4.11 applies for tower cranes with a luffing jib in a situation where their jibs are in the position of minimum outreach. For the maximum outreach of cranes located on opposite sides of the building, staff must ensure that there can be no uncovered areas within the width of the building under construction, that is

$$R_{max}^{crane\_1} + R_{max}^{crane\_2} > D_3 \qquad (4.10)$$

where

D_3 – is the distance between the axes of the tower cranes' railways mounted *on* opposite sides of the building.

Because of safety regulations, both cranes cannot work simultaneously in the area of the same lateral axis of the building. The whole working front of the building, particularly on a long building, has to be divided into assembly zones (cycles) as when two cranes are positioned on one side of the building. In

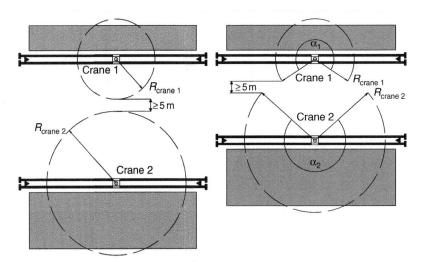

Figure 4.12: Simultaneous work of two cranes positioned between two buildings under construction.

addition, schemes for carrying units to the mounting sites must be determined, coordinate the crane working schedules and the immediate executors of tasks provided with timely and necessary information.

One more option, the simultaneous work of two cranes positioned between two buildings under construction, is depicted in Figure 4.12, where α_1 and α_2 are the restricted slewing angles of corresponding cranes, designed to prevent their jibs crossing.

4.3 Selection and impact areas of mobile cranes

4.3.1 Selection of mobile crane

It is practical to determine the working parameters of mobile cranes using the graphoanalytical method (see Figure 4.13).

The assembly height, that is the maximum required height of the hook H_{max}, is calculated the same way as for the tower crane in Equation (4.1).

$$H_{max} = h_1 + h_2 + h_3 + h_4 \tag{4.11}$$

First the minimum possible length of the boom L_{min} necessary to mount a precast element must be determined, and on that basis the working radius R_{min} and the lifting height H_{min} corresponding to that length are calculated.

Since initially the model of crane is not known, it is presumed that:

$c_7 = 1.5$ m – distance between the centre of the hook and the centre of the cathead axis to hook traverse, lifted to the hoisting height limiter;

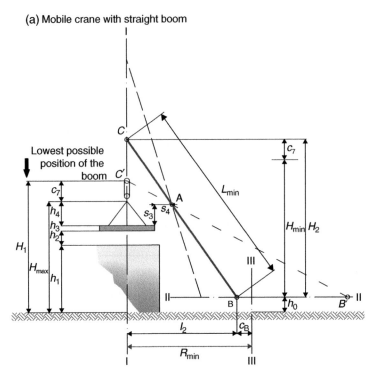

(a) Mobile crane with straight boom

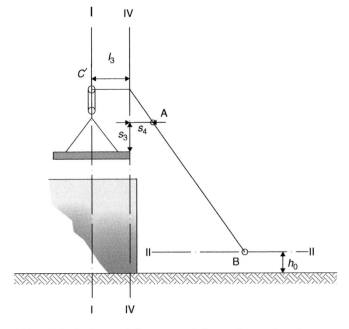

(b) Mobile crane with extended lattice jib

Figure 4.13: Calculating mobile crane minimum boom length.

$h_o = 3.0\,\text{m}$ – height of the centre of the boom heel axis (for mobile cranes it is usually from 1.5 to 3.0 m);

$c_B = 3.0\,\text{m}$ – distance from the centre of boom heel axis to the crane slewing axis (for contemporary cranes it is from 1.8 to 3.3 m). After choosing particular crane c_7, h_o and c_B should be checked and amended if necessary.

To provide the necessary safe distance between the boom and the mounting unit, the safety distances $s_3 = s_4 = 1.0\,\text{m}$ are taken.

Next, the following are drawn:

❑ the mounting unit, at a set scale, at the over lift height of h_2 from mounting level h_1;

❑ assembly axis of mounted unit I-I; and

❑ horizontal axis of crane II-II at height of h_o from the its standing level.

Based on the safety distances s_3 and s_4, point A is calculated, which is the nearest possible point of the crane boom towards the unit.

Next, the lowest possible position of the boom H_1 for this assembly unit is determined:

$$H_1 = H_{\text{max}} + c_7 \qquad (4.12)$$

For the lowest position, the axis of the crane boom is drawn with a flat dotted line $C'AB'$. The minimum necessary boom length L_{min} is found by turning the line $C'AB'$ around point A towards the increase in its slope angle so that one end slides along the assembly axis of element I–I and the other along the horizontal axis II–II. In this way the shortest segment between these axes is detected and drawn by the continuous line CAB, which represents the shortest necessary length of the boom L_{min}. After that it is possible to draw the horizontal projection of the boom l_2.

The working radius of the crane R_{min} corresponding to the minimum length of the boom L_{min} is the horizontal distance from the assembly axis of the unit I – I to the slewing axis of the crane III – III. The slewing axis is calculated by moving right from point B by a distance c_B:

$$R_{min} = l_2 + c_B \qquad (4.13)$$

where l_2 – the horizontal projection of the minimum length of the boom L_{min}, in m.

The lifting height corresponding to the minimum length of the boom is calculated as follows:

$$H_{min} = h_2 + c_7 \qquad (4.14)$$

where H_2 – height of the centre of cathead axis from the standing level of the crane, in m.

If the unit is assembled with an extended lattice jib (see Figure 4.13b), with length l_3, then a new assembly axis IV – IV is selected at the distance of l_3 from the primary assembly axis and is dealt with in the same way as was described earlier with respect to axis I – I.

If a unit is mounted at an angle in relation to the crane's slewing axis, then the measurements of the mounted construction and the assembly unit are increased according to the direction of the crane boom, multiplying them by the value of $1/\cos \beta$:

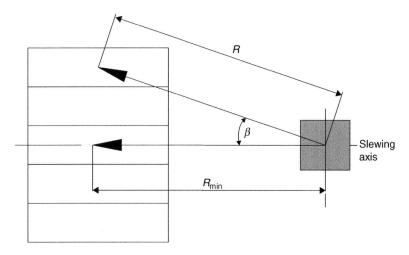

Figure 4.14:   Assembling at an angle.

$$R = \frac{R_{min}}{\cos\beta} \qquad (4.15)$$

where

β – represents the horizontal angle between the crane's slewing
axis and the boom axis in the crane's mounting position
(see Figure 4.14).

When making the decision about a specific crane, a check should
be made as to whether the calculated boom length, radius and
lifting capacity are sufficient to mount the unit respective to the
chosen assembly scheme of the crane. When using a mobile
crane with varying boom lengths, the chart of lifting parameters
varies by each of the various boom lengths.

In Table 4.2, there is an example of calculating assembly parameters of mountable units for guidance when choosing a suitable

Table 4.2 Assembly parameters of precast elements

No.	Precast concrete element	Assembly weight (t)			Assembly height (m)					Assembly radius (m)
		Element	Load take up device	Total	Mounting height	Over lifting	Element	Load take up device	Total	
		g_1	g_2	G_{max}	h_1	h_2	h_3	h_4	H_{max}	R_{max}
1	2	3	4	5	6	7	8	9	10	11
1	Column	11.2	0.2	11.4	0.0	0.5	11.9	1.0	13.4	6.0
2	Frame	15.0	0.6	15.6	10.8	0.5	3.3	3.6	18.2	5.0
3	Roof slab	2.7	0.2	2.9	14.1	0.5	0.3	2.0	16.9	14.5

mobile crane. This is done in similar fashion to the tower crane, as shown in Equations (4.1) and (4.2). The required lifting capacity and lifting height of the mobile crane is determined for placement of various weights of precast elements and for most difficult lifting conditions of a crane. For this purpose, the most heavy and furthest elements from the crane standing position are chosen, and their assembly parameters are calculated, as shown in Table 4.2. The assembly radius R_{max} is determined from the working scheme chosen, that is the sequence of mounting elements and the number of units planned to be lifted from the same standing position, etc.

Based on the assembly parameters in Table 4.2, at least two technically suitable cranes are chosen in order to make a reasonable decision. The technical data for the chosen cranes is recorded, as shown in Table 4.3 (columns 1–6).

The working parameters of crane then have to be set according to the assembly scheme chosen and compared with the calculated assembly parameters of the precast elements (see Table 4.3 columns 7–9). This can be accomplished by comparing the crane lifting charts with the required assembly parameters. The lifting charts can be presented differently by different crane producers (see Figure 4.15 and Figure 4.16).

In Figure 4.15, a lifting chart for crawler crane RDK 25 is presented. Following the chart shows that, for instance, the working radius when lifting a column is 6 m and the respective lifting capacity is 15 t, which is more than required (15 > 11.4 t). From this working radius, the chosen crane is able to lift to a height of 22 m, which also exceeds the required 13.4 m. From this we can conclude that this crane is sufficient for lifting this particular column from the chosen working radius. A similar exercise is completed for all other elements in the table.

Table 4.3 Lifting parameters of chosen mobile cranes compared to the assembly parameters of precast elements

Model of mobile crane	Technical parameters				Units lifted by the crane	Selected working parameters of the crane compared to the assembly parameters		
	Length of the boom (m)	Radius (m)	Lifting capacity (t)	Lifting height (m)		Working radius (m)	Lifting capacity vs. assembly weight (t)	Lifting height vs. assembly height (m)
	L	$\dfrac{R_{min}}{R_{max}}$	$\dfrac{\text{for } R_{min}}{\text{for } R_{max}}$	$\dfrac{\text{for } R_{min}}{\text{for } R_{max}}$		R	G vs.G_{max}	G vs.G_{max}
1	2	3	4	5	6	7	8	9
Option 1								
Crawler crane RDK-25 (with extended jib)	22.5 (main boom)	5/18	18/2	22/15.3	Columns	6.0	15 > 11.4	22 > 13.4
					Frames	5.0	18 > 15.6	22 > 18.2
	5 (extended jib)	10/24	5/1.5	24.8/14.2	Roof slabs	14.5	4 > 2.9	22.5 > 16.9
Option 2								
Mobile crane Liebherr LTM 1030	19.6	3.5/16	17.3/3.7	19/7	Columns	6.0	16 > 11.4	18 > 13.4
					Frames	5.0	17.3 > 15.6	19 > 18.2
	24.8	4/22	13/2.1	24/5	Roof slabs	14.5	4.5 > 2.9	19 < 16.9

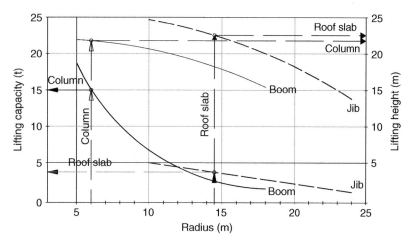

Figure 4.15: Example of determining the assembly parameters based on lifting capacity chart for the RDK 25 crawler crane.

Figure 4.16 shows the lifting charts for the Liebherr LTM 1030 mobile crane, in which the lifting capacity is calculated with the help of capacity table and lifting height with the help of the height chart.

4.3.2 Work of the mobile crane by a recess

When mounting mobile cranes in proximity to recesses and trenches with unsupported slopes (Figure 4.17), the same considerations must be adopted as in the case of tower cranes.

The movement, positioning and operation of construction machinery in proximity to recesses, trenches and holes without extra support is allowed only at a distance determined in the plan of construction works, and must to be outside the margins of the recess slope collapse prism. The positioning of the crane can depend on the depth of the recess and the soil as shown in Figure 4.18.

When working without outriggers or when an outrigger is farther from the recess edge than the crane axis, the minimum distance

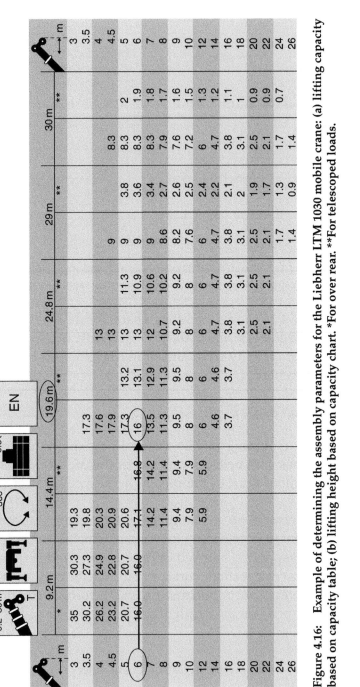

Figure 4.16: Example of determining the assembly parameters for the Liebherr LTM 1030 mobile crane: (a) lifting capacity based on capacity table; (b) lifting height based on capacity chart. *For over rear. **For telescoped loads.

(b)

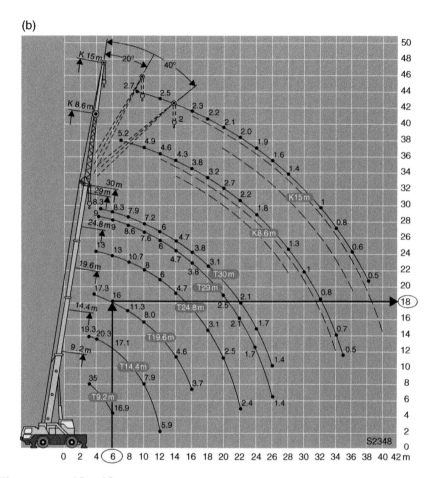

Figure 4.16: (*Cont'd*).

s_5 is taken from the crane axis nearest to the edge of the recess bottom, or from the edge of the crawler track. If working with outriggers, the distance is taken from the centre of the outrigger.

4.3.3 Mobile crane impact range

The impact range of the mobile crane is determined, as in case of the tower crane, using a radius proportionate to the reach of

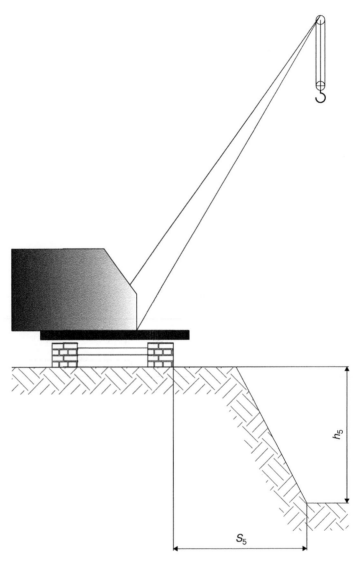

Figure 4.17: Positioning of mobile cranes at the edge of unsupported recess slopes.

the boom necessary for crane works, indicating the slewing restrictions of the boom if required. In contrast to the tower crane, this is done for every assembly position separately (or only for the outermost positions).

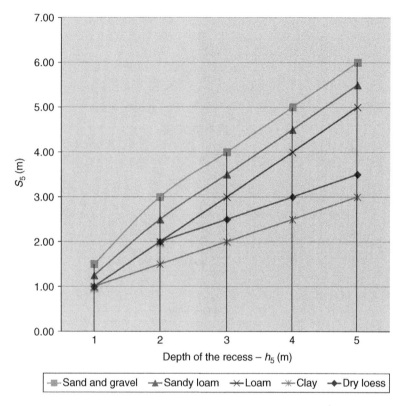

Figure 4.18: The minimal acceptable horizontal distance s_5 from the bottom edge of a recess with an unsupported slope to the nearest outrigger of the crane (m).

For mobile cranes equipped with a boom fall prevention device (Figure 4.19), the distance of the danger area R_4 for mobile cranes is determined by the equation:

$$R_4 = R_{max} + 0.5l_{max} + s_2 \qquad (4.16)$$

where

s_2 – is the dispersion safety distance of possible unit falling. When lifting to a height of up to 10 m, the safety distance is

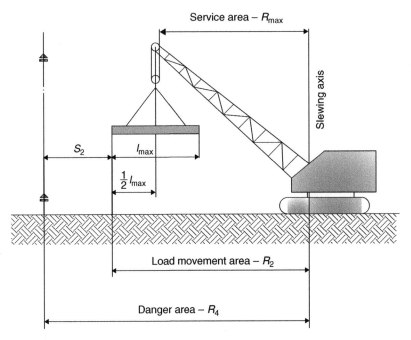

Figure 4.19: Danger area of mobile crane equipped with boom fall prevention device.

taken $s_2 = 0.3H + 1$ m. When lifting higher than 10 m, it is calculated similarly to s_2 for tower cranes as shown in Figure 4.7.

The width of the danger area of the slewing base of a lifting device (as with an excavator) is the sum of the radius of the slewing part and the safety distance (1 m).

If the manufacturer has not given higher safety requirements in the technical documentation for the machine, the safety distance within the working area of the device will be taken as being 5 m from its moving parts and appliances.

The risk area within the working range of a mobile crane is marked by prefabricated removable barriers, flags strung between or by red and white striped signal tape. This marked area for mobile cranes is similar to the risk area of the tower crane's railway.

4.4   Cranes working near overhead power lines

When planning construction works in areas where miscellaneous lifting and assembly equipment has to be used close to overhead power lines, the surveillance and danger areas must be identified.

The protection zone of an aerial power line is in the form of areal space bounded on both sides by imaginary vertical lines along the areal line axes (see Figure 4.20), the range of which depends on the voltage of the line as shown indicatively in Figure 4.21. See also the Electrical Safety Law.

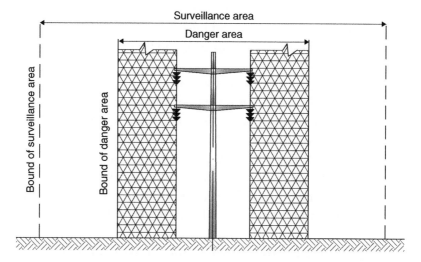

Figure 4.20:   Surveillance and danger areas of aerial power lines.

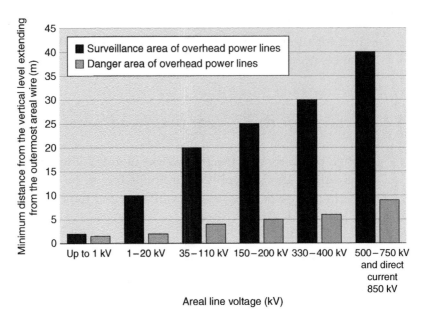

Figure 4.21: Extent of the surveillance and danger area of the electrical overhead power line.

Store of construction materials and use of lifting machinery are prohibited in the surveillance area without the agreement of the organisation that controls the line.

Construction work in live overhead line surveillance area is acceptable only with written authorisation from the organisation that controls the line. This means that this is not only a permission but exact instructions of how works should be arranged. The construction works have to be conducted under direct supervision of a white-collar worker responsible for job safety. The work order/permit is issued with regard to specific work content and with a fixed expiry date.

The work order/permit for construction works in the surveillance area of live overhead line must be signed by the managing director and the person responsible for electrical job safety. The work order/permit is drafted in two copies. One copy is given to the crane operator and the other to the person responsible for job safety (foreman, supervisor, etc.). If construction works are executed on the territory of an operating company, the work order must also have the signature of the person responsible from this company.

Before starting works in the surveillance area, the power must be disconnected from the overhead lines if possible. If it is impossible to disconnect the power, construction machinery can only work with the order/permit within the bounds of the danger area, indicative distances of which are shown in Figure 4.21.

When voltage is 110 kV or higher, construction machinery may only work under live overhead lines if the distance between any outlying part or removable unit of construction machinery and the lowest part of the overhead line is not smaller than the distances of surveillance area (see Figure 4.21).

According to regulations governing the safe use of lifting equipment, cranes (their outriggers or removable units) may not work, or be positioned, within 30 m of the outermost wire of overhead lines without a work order/permit defining safe working conditions.

If it is not possible to adhere to the minimum distances set by the danger area due to construction requirements, the crane's work in the danger area is only allowed after disconnection of the overhead line. The application for disconnection will be made to the company controlling the power line by the person who composed the work order/permit, indicating the time of disconnection. After obtaining written permission to disconnect the power, the work order/permit of works will be issued.

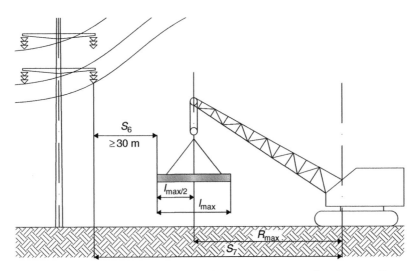

Figure 4.22:   Safe positioning of mobile crane close to overhead power lines.

The safety distance s_7 from the slewing axis of the crane to the nearest outermost wire of the power line (Figure 4.22) is calculated as follows:

$$s_7 = s_6 + 0.5l_{max} + R_{max} \qquad (4.17)$$

where

s_6 – safety distance, but not less than 30 m;
l_{max} – length of the longest unit, in m;
R_{max} – maximum outreach of the mobile crane's boom, in m.

4.5   Hoist danger area

The hoist danger area (see Figure 4.6b) is an area where there is a risk of lifted objects falling to the ground. The width of the area should be at least 5 m calculated from the outer contour of the hoist on the plan. When lifting over a height of

20 m, 1 m is added to the width of the danger area for every additional 15 m. Thus the required width s of the danger area appears as follows:

$$s = 5 + \frac{1}{15(H - 20)} \qquad (4.18)$$

where
H – the lifting height of the load, in m.

4.6   Operating cranes near buildings in use

Operating cranes near buildings bordering the construction site produces a complicated management task that must ensure safety is provided for the people in those buildings, and also for the contiguous pavement and roadways traffic (see Figure 4.23).

Safety requires that the upper ceiling of a building in service must not be in the danger area of the operating crane. If the lower floors of the building are still in the danger area of the crane, the windows facing the construction site must be covered with strong panels (9). The entrance facing the construction site (7) must be closed for the time of construction and taken to the safe side of the building (8).

The construction site fence bordering the building in service should be equipped with a protective screen (10). The minimum width of the passage between the construction site fence and the building in service must be at least 1–1.2 m; in the case of intense pedestrian traffic, this width should be increased.

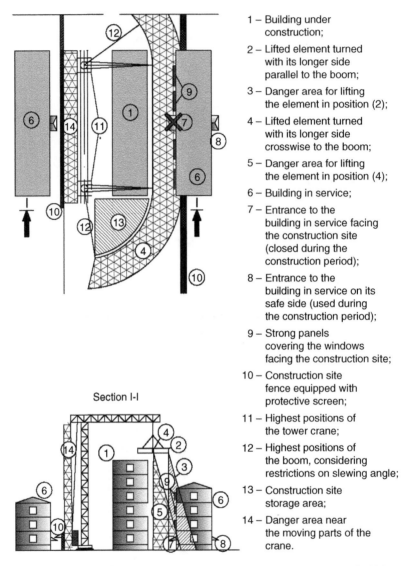

1 – Building under construction;

2 – Lifted element turned with its longer side parallel to the boom;

3 – Danger area for lifting the element in position (2);

4 – Lifted element turned with its longer side crosswise to the boom;

5 – Danger area for lifting the element in position (4);

6 – Building in service;

7 – Entrance to the building in service facing the construction site (closed during the construction period);

8 – Entrance to the building in service on its safe side (used during the construction period);

9 – Strong panels covering the windows facing the construction site;

10 – Construction site fence equipped with protective screen;

11 – Highest positions of the tower crane;

12 – Highest positions of the boom, considering restrictions on slewing angle;

13 – Construction site storage area;

14 – Danger area near the moving parts of the crane.

Section I-I

Figure 4.23: Conditions of operation for tower crane near a building in service.

For the situation presented in Figure 4.23 (see section I-I), it is possible to use the building (6) if the lifted element is turned and held as shown in position 4 in order to reduce the width of the crane's danger area to the required extent (5).

4.7   Restrictions on crane work

When using a tower crane in confined circumstances, there is a need to restrict the movement of the crane, for example the slewing of the jib, outreach of the jib, forward motion of the crane, movement of the load carriage, etc. The restrictions applied are either compulsory or conventional.

Compulsory restrictions are performed by installing sensors and limit switches. These will guarantee, within pre-determined boundaries, the emergency switching off of the crane mechanism irrespective of the crane operator's actions. If several tower cranes operate simultaneously, various automatic safety systems (SMIE's AC30 safety system, Liebherr's ABB system, etc.) that guarantee the safe working of cranes regardless of workers' actions are used.

Conventional restrictions are oriented directly to the attention and experience of the crane operator, slinger or assembler. The reference points for following conventional restrictions are marked on the construction site with clearly visible signs: red flags during daylight and additional red lights or a lantern garland during darkness warn the crane operators of when they are approaching the restricted area. The location of warning signs (reference points) and their design is indicated on the CSL. If they are relocated due to a change of assembly scheme, the crane operators and assemblers will be duly notified.

In order to ensure that conventional restrictions will be followed, the instruction of works management is drawn up for each specific situation. When installing the limiter or the boom's slewing angle, the length of the braking distance of the boom must be kept in mind; for this reason, limiters are installed so that the turning off of the slope would occur 2–3° before the prohibited action zone. If it is desired to limit the movement of the boom by 90°, the limiter should be installed at an angle of 85° (90° − (2 × 2.5°)).

All particular requirements relating to crane operations are drawn onto the CSL, with necessary explanations providing an unambiguous and complete interpretation of the presented solution.

4.8   Working in the danger area

If the limitations due to the dimensions of the construction site or the construction deadline do not allow the safety instructions described in 4.1–4.6, special precautions must be applied:

❑ Issue a work order for high-risk works, appoint a responsible supervisor who stays by the hazardous work at all times.

❑ Draw up work management schemes and work instructions for the crane operator and assembler providing them with timely and proper notification.

❑ Mark the danger areas with visible signal barriers that must be lit during darkness.

Chapter 5
Suggestions for calculating resource requirements

Chapter outline

5.1   Construction site temporary roads

5.2   Construction site storage

    5.2.1   General principles

    5.2.2   Determining the storage space allocation

    5.2.3   Selection of storage locations

5.3   Temporary buildings

5.4   Temporary water supply

5.5   Temporary heating supply

    5.5.1   General principles

    5.5.2   Calculation of heat energy requirements

    5.5.3   Sources of temporary heating supply

The Engineer's Manual of Construction Site Planning, First Edition. Jüri Sutt, Irene Lill
and Olev Müürsepp.
© 2013 John Wiley & Sons, Ltd. Published 2013 by John Wiley & Sons, Ltd.

5.6   Temporary power supply

   5.6.1   General principles

   5.6.2   Calculation of electricity load requirement

5.7   Construction site lighting

5.8   Construction site transport

   5.8.1   General principles

   5.8.2   Calculation of vehicles allocation for car transport

5.9   Load take up devices

5.10 Construction site fencing

5.1   Construction site temporary roads

There must be convenient access and internal roads on the construction site in order to ensure the movement of construction machinery and equipment and the transportation of materials in every season independent of weather. The timely and proper completion of access roads significantly influences the course and costs of construction.

Permanent roads are generally built after levelling of the area and completion of drainage and utility networks. Those permanent roads on the other hand that are usable for transport of construction materials and which do not interfere with overall construction site management may be built earlier together with temporary roads linking them to the unified road network of the construction site.

Temporary roads should preferably be built on the alignment of future permanent roads without laying the last coating. Only if temporary roads lead to temporary storage areas or to buildings away from the alignments of permanent roads should the cost of temporary roads be calculated to the full extent.

The location of roads on the construction site and the traffic scheme must ensure free and safe access for vehicles to the working, assembly, loading and storage areas, and also to the workers' and site managers' rooms. Safer construction site traffic schemes are circular and one-way traffic schemes, which help to prevent vehicle collisions and traffic jams. When planning roads, dead ends that make it difficult for drivers to turn the vehicle around to drive out of the construction site should be avoided. If there are dead ends, a separate roundabout or a road extension of at least $12 \times 12\,m$ must be designed for vehicles to turn around.

It is unacceptable to build temporary road over underground utility networks and in direct proximity to the setting up of utility networks, as this could result in slope collapse and deformation of wearing surface.

The construction site layout must precisely indicate with symbols and explanatory notes the entrance and exit roads, traffic directions, turning places, stopping area for vehicles for unloading and all the linking scales of the planned road units. On construction site with an area of over 5 ha, there must be at least two entrances on each side of the site.

In front of the construction site entrance, a traffic scheme must be installed for vehicles with clearly visible traffic signs (no entrance, limited speed, etc.) at the roadside in accordance with traffic regulations.

The vehicle speed in working area cannot exceed 10 km/h on straight sections and 5 km/h on corners.

Temporary roads have to be built in accordance with the following acceptable minimum distances in m:

❑ between the edge of the road and storage spaces 0.5–1.0;

❑ between the road and standard railway axis 3.75;

❑ between the road and the construction site fence 1.5.

The distance between the edge of the road and the edge of a recess depends on soil type and takes the following requirements into consideration:

❑ for cars and other construction machines with a total mass of

- up to 12 t not less than 1.0 m,

- over 12 t 2.0 m;

❑ the incline of the recess slope must thereby be

- in the case of non-binding or soft surfaces up to 45°.

- in the case of half-binding surfaces up to 60°.

- in the case of rock surfaces up to 80°.

The width of the drive section on a single-lane road is 3.5 m and on a two-lane road 6 m. In case of heavier vehicles (25–30 t or more), the width of the road can increase up to 8 m.

In the case of single-lane traffic, road extensions of up to 6 m are constructed with the length of 12–18 m to ensure passing space for vehicles travelling in opposite directions. Road extensions are also built in the area of loading works, for example in the crane service area. Such passing places are made for at least every 100-m section of road.

The turning radius of the road is selected in accordance with the manoeuvring capability of vehicles, but is not less than 12 m. In curves, the width of the road must be increased to 5 m.

Minimum visibility requirements on the road surface are at least 50 m for single-lane and 30 m for two-lane road. The visibility of oncoming cars should be guaranteed ≤100 m for a single-lane and ≤70 m for a two-lane road.

Structural solutions for temporary roads are classified by the bearing capacity of the subgrade and the loads of the vehicles as follows:

❑ surface-dressed roads;

❑ improved surfaced roads;

❑ hard surface roads;

❑ roads from precast concrete slabs.

The basis for selecting a road type is traffic density, the type and mass of construction machinery and the construction site geological and hydrogeological data. If the bearing capacity and hydrogeological condition of the soil are good, then surface-dressed roads are generally built on smaller sites. Where the soil conditions are more complex, the dirt roads are reinforced by one or two layers of compacted crushed stone, gravel, slag, etc.

Construction site roads with high usage density (dead load ≥12 t) are best built from precast concrete slabs on a 10 to 25 cm thick sand sub-base. Furthermore, it must be borne in mind that prestressed concrete surface slabs can be used three or four times against only one or two times for normal concrete slabs.

If a temporary road crosses a railway, boarding with a counter rail has to be installed in the crossing area and the counter rail has to be installed at the same level as the head of the rail.

Surface-dressed roads are used for one-lane roads with a traffic density of up to three cars per hour where there are well-drained soils (see Figure 5.1a). For precipitation drainage, the wearing surface will be given a cross fall of 2–3% with the help

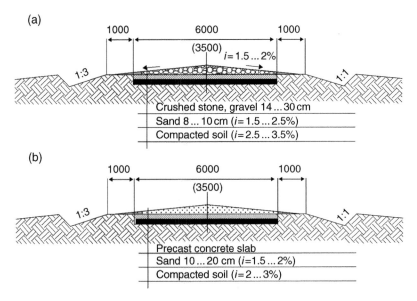

Figure 5.1: Various kinds of construction site road: (a) surface dressed road and (b) road from precast concrete slabs.

of a grader. In the case of heavy loads or an unfavourable sub-base, the road is reinforced with a profiled macadam, gravel or slag covering. The laid coverings are compacted by rolling. In the case of intense car traffic, the reinforced macadam, gravel or slag covering is laid on a sand sub-base compacted with heavy rollers beforehand.

When using heavy vehicles and construction machinery (dead load ≥12 t), the cover slabs are installed on the sand sub-base compacted in the trough recess (see Figure 5.1b).

5.2   Construction site storage

5.2.1   General principles

Storage is built on the construction site for the temporary storage of construction materials, products, construction units and equipment. The main construction materials – gravel, bricks, concrete elements, etc. – are stored in open storage areas. Stores of materials in the storehouses should be as small as possible while still being enough to ensure uninterrupted work.

Construction site storage can be divided into open, closed and half-open storage. Open storage is intended for those materials that do not need protection from weather, such as gravel, concrete and precast concrete elements, bricks, ceramic pipes, etc.

Open storage is mainly located within the range of the tower crane to avoid the need for separate conveyance of assembly units. Only in exceptional cases due to construction site restrictions the precast elements can be stored outside the range of assembly crane.

Closed storage is used to keep expensive materials and materials that perish in the open air (cement, plaster, nails, working clothes, etc.). Closed storage can be heatable or non-heatable. During construction inventory, temporary buildings are broadly used as closed storage. On the basis of transportability and construction, closed storage can be classified as follows:

❑ segment type (prefabricated);

❑ container type; and

❑ trailer storage.

Half open storage – pents – are built to store materials that need protection from either direct sunshine or rain, such as carpentry, soft roofing felt, etc. Pents are located either in the service area of the assembly crane to facilitate the use of crane during loading/lifting of materials, or in proximity of the range of the crane.

Roads must be laid to the storage areas. When storing assembly units within the working range of the crane, the stacking sites for various units must be selected so that, in order to convey the units to the planned position, the crane would have to move as little as possible and make a minimal number of boom turns. For that purpose, units of the same type should be stored at various sites beside the building under construction. The heavier elements and the most frequently used materials must be stored closer to the crane.

Requirements for storage of assembly units:

1) Precast elements must be stored in the technological order of assembly as close to the mounting site as possible.

2) Precast elements must be piled so that from the point of transit and passage the markings are visible and the lifting eyes upwards.

3) Piles should be provided with labels where the type and quantity of the components are indicated.

4) Elements must be stored in conditions that prevent their deformation and soiling and avoid damage to the final surface.

5) The dimensions of piles (height, width) must be calculated according to their specification.

6) Precast concrete units must be piled in a position that corresponds to their load position in building to prevent their breaking under the stress of dead weight.

In principle, the construction site should be provided with materials according to the construction schedule. Nowadays there is no need to store large amount of materials on site in urban conditions; smart and flexible planning is preferred instead. However, possible delays should be considered and for that reason there should be space foreseen and indicated on construction site layout for storing reasonable amount of construction materials.

The situation is different in case of building in unsettled regions. For these cases, the storage area should be calculated by types of materials and drawn on construction site layout.

Materials sensible for moist and other weather conditions should be placed under pents.

Precast concrete elements, construction blocks, bricks and lumber can be piled in open storage area. Needed space should be calculated according to their specification. Large precast elements should be placed as close to their working position as possible in order to prevent multiple unnecessary liftings from one place to another.

The method of piling the elements should ensure stability of the piles and convenience in lifting the units.

Passages in between the piles have to be between every two piles along the length of the building and not less than every 25 m across the building. In between the piles along the building every 15–20 m, there should be transit points of at least 1 m in width to ensure free passage.

If the storage site is adjacent to a recess, the locations of the piles must be planned outside the borders of the collapsing prism of an unsupported slope. If the assembly units are stored nearer the recess, a control calculation for the slope stability is performed, taking into account the dynamic loads, and slopes will be buttressed if necessary.

For operations such as 'assembly on wheels' the units are transported straight to the assembly area and taken directly from the cargo vehicle to the mounting site; only small components are stored in the site storage.

The sites for unloading from vehicles, and the vehicle roads, are added to the construction site layout with the intention that the crane need not change the outreach of the lifting hook and fly-jib when conveying components to their planned positions.

5.2.2 Determining the storage space allocation

In determining the necessary size of storage space, the nomenclature of products and materials stored, and their method and conditions of storage, must be taken into account.

When calculating the space necessary for storage, the quantity of materials required per day and labour intensity, from which the maximum daily requirements is derived, must be considered.

Estimated materials reserve M_0 is determined as follows:

$$M_0 = \frac{M \times k_1 \times T_m}{t_i} \qquad (5.1)$$

where

M – the total need for materials per accounting period t_i;

t_i – duration of accounting period in days;

T_m – standard of material reserve in days; taken experientially on the basis of the data of the construction company. This depends on the location of the construction site and material providers, intensity of construction schedule, risk bearing in case of delay, etc.

k_1 – coefficient for uneven usage of materials determined by construction company, taken as roughly 1.3;

Useful storage space (without passages or transit roads) is calculated as:

$$S_m = \frac{M_0}{M_1} \qquad (5.2)$$

where

M_1 – is the amount of material that is possible to store on $1\,m^2$ (see recommendations in Table 5.1).

The overall area of the storage sites consists of three components:

❑ space under the materials, units and constructions;

❑ space necessary for receipt and hand over of materials;

❑ space necessary for passages and transits;

and is expressed as:

$$S_s = k_2 \times S_m \qquad (5.3)$$

Table 5.1:   Average space required for storage of construction materials

No.	Material	Measurement unit	Average storage space required (m²) per unit of material
			M_1
1	Steel, reinforcement	t	1.5
2	Timber	m³	1.5
3	Bricks	Thousand pieces	2.5
4	Natural stone, gravel and sand	m³	0.5
5	Pipes	m	2.0
6	Cable	t	5.0
7	Precast concrete elements		
	• Foundation elements	m³	1.5
	• Columns, ceiling slabs	m³	2.0
	• Roof slabs	m³	3.6
	• frames	m³	3.5
	• Beams	m³	5
	• Wall panels	m³	1.0

where

k_2 – is the coefficient counting for passages and transits and is valued in average from 1.2 to 1.4, less for bulk material and more for storage in bins and storehouses.

When planning construction works in relation to a certain project, it is wise to specify the necessary materials reserve (norm) in days depending on the agreed procurement charts and calculate the storage needs according to materials specification for the particular project.

5.2.3   Selection of storage locations

As noted, the locations of storage have to be planned simultaneously and in compliance with the planning of location and assembly schedule of the assembly cranes. If additional roads are not required, the storage sites are planned along the designed alignments of roads together with necessary extensions. Temporary access roads must be built for separately situated storage.

The dimensions of storage sites and the types of units stored must be indicated on the construction site layout. It is not acceptable to stack different types of units into one pile. Receiving sites for mortar and concrete must also be indicated.

The surface of open storage sites have to be planned with an incline of 2–5° in order to ensure the drainage of precipitation. For non-draining soils, a non-watertight layer of ground is laid with a thickness of 5–10 cm.

5.3 Temporary buildings

Temporary buildings are defined as different service and support buildings that ensure a controlled course of construction work on the main building under construction throughout the entire construction period.

Temporary buildings can be divided into production, office, storage, workers' and public buildings according to their function.

Depending on their structural solution, the temporary buildings can be built either for one-time use or they can be prefabricated structures, which are designed for frequent displacement and usage on different construction sites.

From the point of view of mobility and structural peculiarity, temporary buildings are classified as trailer, segment or container type.

Trailer buildings are caravan-type rooms on wheels that are practical for small-scale constructions that involve distances (road constructions, power lines, etc.).

Segment or prefabricated buildings are assembled on site from prefabricated and unified manufactured panels. Exterior wall panels are insulated and have window openings and/or doorways where needed. From these unified panels, it is possible to complete various buildings with desired floor plans and spatial solutions. Bolt joints make the assembly and dismantling of panels easy and fast.

A container-type building is a prefabricated building mounted on a rigid frame from which a building complex with the necessary function, size and floor plan is completed on site. This kind of container building can come in various sizes and various supply levels according to the purpose of the building.

Sanitary rooms for workers can be located in:

❑ prefabricated standard buildings;

❑ the office wing of large sites;

❑ existing rooms of the building if these are available on the construction site.

Temporary sanitary rooms can also be located in the adjusted buildings on construction sites that have buildings due for later demolition. This possibility can help to lower the construction company's site costs.

Sanitary rooms are divided into the following categories according to purpose of use:

❑ dressing rooms;

❑ washing rooms;

❑ showers;

❑ rooms for drying clothes and footwear;

❑ heating and resting rooms;

❑ canteens;

❑ toilets, including women's sanitary rooms, etc.

In positioning temporary buildings, the following principles should be considered:

❑ In between the temporary buildings there should be convenient and safe passages with a reinforced surface of not less than 0.6 m wide;

❑ Temporary buildings cannot interfere with the course of construction work throughout the construction – this requirement applies first of all to non-prefabricated buildings;

❑ Buildings should be linked to ensure rational and economical connection with utility networks.

Temporary office and workers' buildings must be located:

❑ outside the risk areas of construction machines and vehicles;

❑ on the windward side in relation to objects emitting dust and harmful gases and not closer than 50 m;

❑ near the entrance of the construction site so that there is access to the dressing room and from there on to the street without crossing the working area;

❑ at a distance of at least 24 m from the buildings under construction and any auxiliary buildings.

Recommended distances between temporary buildings are as follows:

❑ between dressing room and place of work ≤ 500 m;

❑ between canteen and place of work ≤ 500 m;

❑ between heating rooms and place of work ≤ 150 m;

❑ between toilets and place of work ≤ 100 m;

❑ between drinking place and place of work ≤ 75 m;

❑ between temporary buildings and construction site fence ≥ 2 m;

Temporary buildings should be set in groups of up to 10 containers with the distance between each group of buildings at least 18 m.

A first-aid post is foreseen on any construction site where over 300 people work. If there are from 150 to 300 workers, the first-aid post has to be included in the supervisor's office as a separate room ≥ 12 m². With less than 150 workers, the supervisor's office must have a first-aid kit.

On the construction site layout, the overall dimensions of buildings, the linking of buildings and utility networks, construction site passages as well and access roads must be indicated.

In the explication of temporary buildings and facilities, the number and name of each should be indicated in terms of their cubage (m³), area (m²), trade mark or structural solution.

5.4   Temporary water supply

Planning the temporary water supply takes place in the following order:

❑ calculating the water requirement;

❑ selection of the supply source;

❑ planning of water conduit scheme and selection of mains materials;

❑ calculation of mains dimensions;

❑ linking the water network with consumers on the construction site layout.

During the construction period, the need for water Q_w (l/s) is summed up from three components:

❑ production water;

❑ general water; and

❑ fire water.

In principle, it is possible to calculate these separately but considering that fire water is most crucial and essential between these we can simplify the calculations by assuming that if the need in fire water is guaranteed then it will cover the water capacity for production and general needs also.

The minimum requirement for fire water is determined by the consumption of two water spouts feeding simultaneously from a hydrant at a rate of 5 l/spout. Hence $Q_w = 10 l/s$ for construction sites with an area of up to 10 ha and $Q_w = 20 l/s$, for construction sites of up to 50 ha.

The conduit diameter for fire water must be at least 100 mm. The alignment of the temporary water conduit and the locations of fire hydrants (maximum distance 100 m from potential fire) have to be indicated on the construction site layout.

If it is planned to use natural bodies of water as fire water sources, then proper hydrants and access roads for vehicles have to be built and clearly visible signs of locations, distances and traffic scheme to the hydrants must be installed on the construction site.

In addition, the construction site has to be equipped with fire extinguishers according to fire regulations as well as a fire protection cabinet (including axes, crowbar, shovels and a gaff) and a sandbox (at least $0.5 \, m^3$), etc.

In all cases, the need in fire water has to be verified with local fire and safety regulations.

5.5 Temporary heating supply

5.5.1 General principles

A temporary heating supply to the construction site is necessary to:

❏ supply technological processes with heat energy, for example for heating water and aggregates in concrete and mixture nodes, to heat shelters and concrete and to defrost soil, etc.;

❑ dry and heat the buildings under construction;

❑ ventilate and supply heat to temporary offices and workers' buildings (dressing rooms, showers, canteens, rooms for drying clothes, etc.).

Temporary heat supply systems are used during construction work and are dismantled thereafter.

Temporary heat supply systems consist of the following components:

❑ sources of heat energy;

❑ temporary heating systems,

❑ terminal equipment, such as heaters, water boilers, heat blowers, etc.

The design of a temporary heating supply for construction includes the following:

1) Total requirement of heating energy for the construction object or complex is calculated separately for all consumers.

2) Sources of heating energy are determined and fuel consumption is calculated;

3) Heating piping is dimensioned and designed.

4) Local heating and drying units, steam generators, etc., are selected.

5.5.2   Calculation of heat energy requirements

The heat energy required for technological use and works during periods of frost are calculated according to construction work technology standards.

The overall heat energy requirement Q_h (kJ/h) of the construction site consists of the following energy components:

❑ energy required to heat buildings and shelters. This depends on the cubage of the building (m³), special heat coefficient (kJ/m³ h °C) and outdoor and indoor temperatures of heated buildings. In average, the outdoor temperature influences the energy requirements around 10–20%. The special heat coefficient is taken in accordance with local construction regulations and is from 2 to 5 kJ/m³ h °C;

❑ energy required for the drying of buildings. In order to determine the quantity of air and heat energy necessary to dry the building, extra calculations are required, including calculations of the heat energy needed to vaporise moisture from structures and heat the air in the building.;

❑ energy required for technological purposes.

While calculating the overall heat requirement, the heating conduit losses – approximately 15% – should also be taken into account.

5.5.3   Sources of temporary heating supply

Sources of temporary heating supply can be:

❑ existing or designed heating systems that connect the construction site to an existing district/company boiler house; or

❑ a temporary boiler house;

A temporary boiler house is used when an existing source of heat energy is absent or the source lacks available calorific power. Such a situation can occur prior to building handover when intensive drying requires a lot of heat energy.

The necessary heating surface of a temporary boiler is calculated according to the overall heating requirement Q_h and heat capacity output of the boiler in kJ/m^2h (according to equipment documentation).

Temporary heating units can operate on gas, liquid fuel or coal as well as electricity. Heat carriers can be steam, water, air or a mixture of gas and air and radiant energy. Buildings that house temporary boilers are either prefabricated (segment type), container or trailer type. Small heating units can be set into a heatable building if they do not interfere with construction work.

Lately there has been extensive use of heating units where the heat carrier is air, that is heating ventilating units. Room ventilation significantly speeds up the airing and drying of constructions, which is important when finishing works occur in winter or spring.

Electric calorifiers (hot air blowers) are the most convenient heating devices, although because of the relatively high price of electricity, their expediency has to be justified economically. When connecting electric calorifiers to the mains, it is vital to pay special attention to the electrical safety regulations and avoid overloading power lines.

It is practical to install heating calorifiers in a large room (workshop, hall, etc.) or next to stairways of dwellings. Calorifiers are manufactured with various outputs and in various complexities. That allows a suitable and economically acceptable technical solution to be found for every object. Unlike other heating devices, air heating devices don't require supervision from personnel and provide a continuously functioning temperature regime in the heated rooms. For vertical distribution of hot air, canvas sleeves are used in dwellings, chutes and vents providing them with special tube connections.

Air heaters with heat exchangers are used to heat and dry buildings as the main heating devices and as an additional heat source during finishing works.

Heat generators are the main heat sources for the outdoor work of soil defrosting, bitumen heating, etc. For the heating of rooms they are used as additional heat sources; these generators use heating oil for fuel and also mains and bottled gas.

Infra-red radiators, using bottled gas, are mainly for drying various structures and elements. The infra-red part of the spectrum permeates the comparatively small air layer between the radiator and the heatable surface almost without loss and heats the radiatable surface regardless of the temperature of the surroundings.

Steam generators are suitable for outdoor works in winter, including melting frozen soil and snow and defrosting frozen water pipes.

Temporary heating systems are designed as single-end schemes, with the conduits placed into a trench. The pipes are insulated with milled peat, slag or light gravel, which also gives protection from moisture.

5.6 Temporary power supply

5.6.1 General principles

The provision of electricity to the construction site is an important precondition to ensure the normal course of construction works. After the growth of the industrialisation of construction and the mechanisation of works, the importance of an electrical power supply to the construction site has grown significantly as the whole electrical economy has become more sophisticated.

Today the annual consumption of electricity is calculated as over 4000 kWh/worker. This is why the planning of construction site electricity can be considered one of the main assignments in construction site management.

The planning of construction site electrical supply has the following general requirements:

❑ supplying construction with electricity in the required quantity and quality (voltage, frequency);

❑ flexibility of the electrical system – the possibility to supply all consumers at every place on the construction site with electricity;

❑ reliability of electrical supply;

❑ supply of the required level of consumption with minimum network losses.

In planning the construction site electrical supply, the starting point must be to determine the location of the consumers on the site and their power requirements. The possible power sources are then identified, taking into account the requirements and restrictions that occur with a moving work front. In supplying the construction site with electricity, permanent sources and objects of electrical supply (substations, cable lines, etc.) must be used as much as possible.

The order of planning for the supply of temporary electricity to the construction site is as follows:

❑ calculating required consumer power;

❑ determining the number and output of transformer substations and other electrical supply points;

❑ determining which objects need additional electricity, for example repelling water, heating concrete, etc.

❑ determining the location of transformer substations, distributive networks (power and lighting lines) and switchgear (main switchboards and distribution boards) on the CSL;

❑ plotting the electricity supply network scheme and determining the required technical parameters.

When calculating electrical load, the construction site layout, time schedule of works, description of construction works, parameters of construction machinery and mechanisms and the building's technical engineering data are used as initial data.

Generally, alternating current with a frequency of 50 Hz and a voltage of 380/220 V is used on the construction site – for engine installations 380 V and for lighting 220 or 36/12 V.

When planning temporary electrical supply for the construction site – which has to be built during the preparation stage – existing or planned power facilities (low- and high-voltage lines, transformer substations, thermal power stations) must be intended as power sources as much as possible. For the supply of structural objects, step-down transformer substations of up to 0.4 kV must be used. On large construction sites, and in the case of large energy requirements, several substations can be used with voltage being stepped-down to 6 kV first and subsequently to 0.4 kV.

When placing line and other objects in places where the use of existing power lines is impossible, transportable electric power stations or machines with electricity generators (welders, etc.) must be employed. This may also be applicable if the contractor has not managed to connect to an existing power line by the beginning of the preparation stage.

Power lines for electricity consumers on site are connected to 380/220 W or 220/127 W substation output boards. For construction objects that cannot be supplied through standing substations, transportable unit substations connected to high-voltage aerial and cable lines must be provided. Local consumers of electricity are connected to the construction site through distribution and group boards.

The distance between the consumer of electricity and the power source (transformer substations, unit substations, transportable electric power stations, etc.) should not exceed 200–250 m to prevent a large voltage drop in power lines and the resulting power loss and disturbance in device operation. However, to prevent these deficiencies, an increase in wire diameter cannot be considered economically practical.

Consumers are linked to the power source either with a dead-ended feeder or a ring network – a mixed system. In selecting the connection scheme, stability of the output necessary for the consumer of electric supply and the cost of setting up supply lines is taken into consideration. Consumers with low needs are mainly supplied through dead-ended feeders, while consumers with greater needs are supplied through a ring network. In the case of higher requirements for stability of electrical supply, for example during caisson works, where the object would have to be supplied with several power sources, a back-up power station might be linked to the system. Electricity lines are built as cable or overhead lines.

5.6.2 Calculation of electricity load requirement

Several methods can be used to determine the estimated electricity load depending on the characteristics of the initial data as well as the desired level of detail and precision of the calculation.

Estimated construction electricity load Q_e (kWA) through particular electrical charge energy is calculated as follows:

$$Q_e = \frac{\sum Q_i \times W}{T_{max} \times \cos \phi}(kWA) \qquad (5.4)$$

where

Q_i – particular charge of electric energy for each kind of work or production unit (taken from reference books);

W – annual amount of work or production in physical indices;

T_{max} – number of working hours in a year according to intensity of work (man-hours/year);

$\cos \varphi$ – output factor, the value of which depends on the number of machines and their load (in average 0.5–0.8 for cranes and machinery and 1.0 for lighting, is taken from reference books).

It is possible to estimate the electricity load Q_e on the basis of installed output:

❑ undivided by individual consumers, based on total electrical output installed or

❑ divided into consumer groups, based on output of the load consumer, output for technological needs and output of outdoor and indoor lighting equipment. While calculating the electricity load, the network losses from 5% to 10% but also possible demand-side factors depending on particular equipment have to be considered. The values are taken from reference books and catalogues.

The output needed for outdoor lighting can be calculated from the luminosity (lux) of the surface or the electrical charge for lighting of $1\,m^2$ of surface according to local construction regulations. In average, the recommendations as shown in Table 5.2 could be used.

On the basis of the acquired data, the electricity load chart for the construction site is formed and the time of peak load is determined, specifying both the list of construction machinery and their technical indices. The output of the selected power sources (transformer substations, etc.) has to cover the required total power output during construction peak loads.

Table 5.2:   Recommendations for surface lighting in construction

Appliance	Average illumination (lux)
General lighting of the construction site	2
Passages and thoroughfares	25
Mural and assembly works	20
Storage and loading works	10
Finishing works	50

5.7   Construction site lighting

The planning of construction site lighting includes determining the necessary level of lighting in various areas of the site, selection and location of lighting equipment (type, individual output), calculation of the required power output, and planning of the mains lines.

Construction site lighting is divided into work, emergency and surveillance lighting. With work lighting, general and local lighting is distinguished. On the construction site and in the work area (job site), there should be even general lighting, and where required for better visibility there should also be additional local lighting. The lighting should suit the nature of work (see Table 5.2).

Emergency lighting is built on independent supply and installed mainly in passages and slopes with a luminosity of not less than 0.2 lux. Lighting of surveillance area begins from 0.5 lux.

Lights are installed either on existing structures, on stationary and transportable poles or supports and on natural ridges.

The planning and realisation of outdoor lighting is made more difficult by the changing construction site and working levels in time and space, which obliges relocation of lighting equipment. In such cases, mobile lighting equipment should be preferred, for example trailer masts on rubber wheels or rail track; wood, steel lattice frame or telescopic constructions can be used as masts.

When lighting a construction site, special attention must be given to reducing the number of lighting points, ensuring at the same time the proper lighting of the territory used, especially the places of work; the reliability of the whole lighting system must be ensured at a reasonable level of cost.

Construction site lighting is designed at the planning stage of construction site management. Electrical installation work is usually carried out with a specialised unit. This could be a company that has the necessary material and manpower and which completes the whole work cycle starting from planning and maintenance and finishing with the dismantling of the system. The company also ensures the reliability and safety of the lighting.

The calculation of the required number of floodlights, n_f, is made on the basis of the output of the light sources, q_4 (W/m² lux), the luminosity of the surface, E (lux), and the estimated size of the lighted surface, S_1 (m²), according to the following equation:

$$n_f = \frac{q_4 \times E \times S_1}{q_i} \tag{5.5}$$

where
q_i – the output of each incandescent lamp W.

In damp rooms, it is advisable to use voltage reduced by 36 V. Voltage reduction takes place on a special switchboard.

5.8   Construction site transport

5.8.1   General principles

Traffic is an important part of the continuous construction flow, connecting construction sites to factories, quarries, storage areas and other sources of the required resources. Expenses on transport and loading constitute a relatively large portion of construction materials costs (up to 10%).

Construction transport is classified in two groups: the inner construction site (so-called technological) and the outer construction site, according to the direction into horizontal, vertical and incline transportation.

Practically all kinds of transport are used in construction:

❑ car transport;

❑ railway transport;

❑ waterborne transport;

❑ tractor transport;

❑ air transport;

❑ pneumatic pipeline transport.

The main kind of transport on construction sites however is car transport, which has the predominant proportion. A decisive advantage of car transport is its mobility and manoeuvring ability, the ability to transport the material directly to the site of consumption and to a certain extent its self-loading ability.

Rail transport is justified on construction site only in case the large production complex is to have a standing branch line, the relief of the construction site is even and the size of the load is 400 000–500 000 t/year.

Tractors are used on construction sites mainly where there are difficult road conditions and complex relief, and where heavy construction units must be delivered to the assembly site across relatively short distances.

Air transport is used on construction relatively seldom, mainly for transporting people and materials to regions that are difficult to access – for example islands where ships do not sail – for operational movement of equipment and materials, or during installation of equipment in high buildings, for example installation of antennae on top of high buildings, etc.

Compressed air is used mainly when loading cement onto a cement transporter, removing sawdust, etc.

5.8.2   Calculation of vehicles allocation for car transport

Construction traffic is determined by the volume, type, freight turnover, and freight, as well as the possibilities for organising flow.

Load intensity is the volume of construction materials and structures transported per unit of time, in tonnes. Freight turnover is the amount of transport work per unit of time, in tonne-kilometres. Commodity flow is calculated as part of freight flow in one direction.

The basis of calculations is the time schedule of construction works and the quantities of necessary materials expressed by time interval. Flow rationale variants for vehicle use are formed from freight turnover and commodity data.

The daily allocation of vehicles (in number) is calculated on the basis of overnight commodity flows on certain routes, from which the transport schedules are compiled. When calculating, technical (correspondences between the nature of the transportable goods and vehicle parameters) and other conditions (deadlines, road conditions, etc.), as well as economic considerations, must be borne in mind.

In evaluating vehicle variants, cost efficiencies must form the basis of choice:

$$C_t = C + \sum K + \mu \qquad (5.6)$$

where
C – cost price of load delivery, in €;
K – cost of capital assets (vehicles), in €;
μ – efficiency coefficient of capital assets.

Cost price of construction site loads is determined from:

$$C = C_1 + C_2 + C_3 \qquad (5.7)$$

where
C_1 – operational costs of transport-related buildings during period in question, in €;
C_2 – cost of loading and unloading, in €;
C_3 – operational costs of vehicles, in €.

Vehicle allocation per shift/day is found using:

$$N_t = 1.1 \times \frac{G_{24}}{P} \qquad (5.8)$$

where
G_{24} – 24-h load volumes
P – vehicle productivity/24h; this is calculated from the vehicle carrying capacity, a factor of carrying capacity utilisation, haul distance (km), driving speed (km/h), and a factor of transit usage and the period of loading and unloading during the haul cycle.

5.9   Load take up devices

In lifting, transporting and mounting construction units into their planned positions, and also in loading and unloading and other crane operations, load take up devices are used to strop and/or hook units.

Load take up devices have to be universal, simple and light. The low weight of the device is significant because the mass of the take up device has a direct influence on the selection of crane (which is based on lifting capacity).

Steel constructions, precast concrete elements and other units are lifted with the help of slings, cross beams (traverses) or grips. Attaching the liftable unit to the lifting hook of the crane – stropping – is one of the more responsible operations of the mounting works complex. This is why the main condition for slings is their reliability, complete safety and simplicity of use.

According to the structural solution, slings come under the heading of simple load take up devices. Cable slings are extensively used in construction work and are divided into:

❑ single branched;

❑ double branched;

❑ triple branched;

❑ quadruple branched;

❑ double lifting eye;

❑ ring slings.

Normally, the producers provide relevant data along with their products. However, when calculating the lifting capacity of steel cable slings, the basis used is the number of sling branches and their angle of inclination towards the vertical; accordingly:

$$F = \frac{Q}{n \times \cos \beta} (\text{N}) \tag{5.9}$$

where
F – internal force inside the sling branch;
n – number of sling branches;
Q – mass of the liftable load;
β – inclination angle of sling branch towards vertical.

The force in the sling branch F depends on the location, and the number and intervals of lifting eyes in the liftable unit. The longer the distance between lifting eyes (A), the greater is the angle α between sling branches of the same length (L_s), and consequently the greater is the force per branch (see Figure 5.2).

In the case of general-purpose multibranched slings, the angle α between the branches is taken to be $\leq 90°$. The optimal

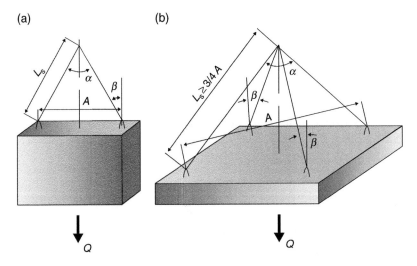

Figure 5.2: **Double- and quadruple-branched slings.**

inclination angle β of a sling branch towards the vertical is 30–40° ($\beta = \alpha/2$).

One can test the suitability of the sling for the load by the relationship between the maximum distance A of lifting eyes and the length of the sling branch L_s.

The length of the sling branch is selected on the basis of the distance between the lifting eyes of the liftable unit so that the angle between the sling branch and the vertical line $\alpha/2$ will not exceed 45° (see Figure 5.2).

Calculating the relationship between the length of the sling branches and the distance between the lifting eyes of the liftable unit is as follows:

❑ If $A/L_s = 0$, then the angle between sling branches $\alpha = 0°$.

❑ If $A/L_s = 1$, then the angle between sling branches $\alpha = 60°$.

❑ If $A/L_s = 1.41$, then the angle between sling branches $\alpha = 90°$.

❑ If $A/L_s = 1.73$, then the angle between sling branches $\alpha = 120°$.

Based on this it can be seen that, if the relationship is $A/L_s \leq 1.41$, that is the angle between the sling branches $\alpha \leq 90°$, then the length L_s of the sling branches is suitable for lifting the construction unit for distance A of the lifting eyes, naturally taking into account the carrying capacity of the sling.

When lifting an assembly unit with one or two lifting eyes, a four branch sling can be used, hooking it with either one or two lifting branches. In this case, the carrying capacity of the sling has to be taken as two or four times smaller than that certified by the manufacturer as the carrying capacity correspondent to the certificate of the sling has been computed for simultaneous use of all branches.

The smoothest and safest take up devices are grips, which can pick up the element (structure) and prevent it from falling during lifting, as well as release the unit after assembly. Grips are more reliable than slings and reduce the amount of manpower used when stropping/hooking. Various types of grips are used during construction, from the simple to the sophisticated. Grips are mainly used for lifting brick packages, poles and wall panels.

Units with large dimensions are mounted using traverses. Unlike slings, traverses have a rigid construction. This enables the reduction of the height of the take up device, forces the sling branches inward and lowers the compressive forces in the liftable construction unit. Panels, stair flights, poles and other structures with shifted centres of gravity are moved and mounted using balancing, so-called, beam traverses.

With balancing traverses, the sling branches are under even load, and it is also more convenient to carry some construction units into their planned position. To ensure the assemblers are working in safety, for example during the mounting of high poles, the load take up devices should be equipped for distance unhooking.

Interchangeable load take up devices (slings, traverses, chains, etc.) are guaranteed by the technical certification of the manufacturer, and after repair to certification issued by the workshop that supplied the service.

During technical certification a visual inspection is conducted on the load take up device; it is then loaded at 1.25 times their rated load capacity for a duration of 10 min.

During use, load take up devices must be inspected regularly using a schedule of:

❑ traverses after every 6 months;

❑ slings (apart from those used very seldom) every 10 days;

❑ grips every month.

Crane lifting hooks must have bolts to avoid the take up device spontaneously unhooking. In addition, it is advisable to supply sling lifting hooks with protective shutters.

In all cases, the frequency of inspection and other requirements have to be in accordance with local construction regulations.

5.10   Construction site fencing

In a populated area, or on the territory of an operating company, the construction site must be surrounded by a fence to prevent outsiders from coming into the construction works area.

Construction site fencing is divided into the following categories according to use:

❑ protective and guard fencing, which is to prevent outsiders from gaining access to the construction site and any areas that present risk, and to prevent theft of materials;

❑ safety fencing, which prevents access to a particular task area for those unconnected with that task, and which prevents risk associated with objects falling from above;

❑ signal fencing, which marks the boundaries of areas with risk factors.

Safety fencing is generally made from inventory units (plates, handrails, protective screen components, etc.). Plates can be made of light or dense boarding or as framed web fence units mounted on special concrete legs.

The height of protection and guard fencing (with and without protective screen) should generally be taken as 2 m; however, the following heights maybe used:

❑ protective fencing (without screen) – 1.6 m;

❑ protective fencing (with screen) – 2 m;

❑ protective fencing of job place – 1.2 m;

❑ signal fencing – 0.8 m.

The lengths of barrier plates are 1.2, 1.6 or 2.0 m. The distance between the columns of the signal fence is up to 6 m. The protective screen is made with a 20° slope towards the construction site, the screen has to reach 50–100 mm above the level of the passage and the construction has to be able to bear at least 16 N of concentrated burden in the centre of the bearing opening. Inventory passages are planned for a 20 MPa standard load and supplied with 1.1 m high handrail and 0.5 m high horizontal intermediary beam. The minimum width of the passage is 1.20 m.

The fixing construction of the plates of the perimeter fence has to enable them to be linked on ground with up to a 10% gradient.

Chapter 6
On-site safety requirements

Chapter outline

6.1 General basics and responsibilities

6.2 The duties of building contractors

6.3 The obligations and rights of the labourer

6.4 Ensuring safety on the construction site

    6.4.1 General

    6.4.2 Safety requirements in a work zone

    6.4.3 Special requirements for assembly works

    6.4.4 Special requirements for work in pits, wells, in tunnels and earthworks and underground

    6.4.5 Special requirements for working at height and on roofs

    6.4.6 Special requirements for demolition work

    6.4.7 Ventilation in the workplace

    6.4.8 Emergency exits from the workplace

The Engineer's Manual of Construction Site Planning, First Edition. Jüri Sutt, Irene Lill and Olev Müürsepp.
© 2013 John Wiley & Sons, Ltd. Published 2013 by John Wiley & Sons, Ltd.

6.1   General basics and responsibilities

The responsibility for the building site, including work safety, lies with the owner of the real estate, as long as he or she has not delegated that responsibility to another organisation through a contract of services or an authorisation agreement.

The owner of the construction site is required to ensure:

❑ the maintenance of the construction and its land units and the safety of the surrounding environment during the construction and exploitation of the building. This includes preventing access to buildings with a danger of collapse or signs of deterioration until they have been demolished or renovated. This has to be done with warning signs unless a contract of services says otherwise;

❑ delivery of the proper notice of construction to the local government (except for small construction) at least three working days prior to the construction (unless a contract of services says otherwise) if:

- the expected duration of the construction exceeds 30 working days and at the same time there are >20 labourers on the construction site, or

- the expected volume of work exceeds 500 staff-days;

❑ the opportunity/access for control to be exercised by national and local authority oversight organisations and building inspections;

If the owner uses a contractor or a professional management company, then the responsibility for work safety lies with them.

For the manufacturing operations on the construction site, a work environment that does not damage the surrounding environment nor endanger the lives, health or property of the labourers or a third party must be created. If construction is in an area of heightened danger, the physical, chemical and other hazard parameters of the work environment cannot surpass the set maximum levels. A maximum level is the average hazard parameter per time unit that does not damage the health of a worker in an 8-h working day (a 40-h working week).

The company handling the supervision of the owner of the construction is required to check:

❑ that work safety and healthcare regulations are met, that the contractor is not polluting the surrounding environment and keeping the construction site properly maintained, and, if need be, making proper entries in the site diary;

❑ that entries in the site diary are actioned.

If construction has a main contractor, prior notice of construction has to be delivered by the main contractor. If there is no main contractor, the owner of the construction must appoint a contractor to be responsible for health and safety on the site and inform other contractors of this fact.

The main contractor has to prepare a list of dangerous operations on the site, guided by the following list of the foremost dangerous operations on a construction site:

1) operations that can cause a landslide or engulfment, and where the danger might be increased by the work methods used or the environment where the construction site is located;

2) operations in which labourer health can be compromised by biological risk factors and dangerous chemicals, including asbestos;

3) operations that are located in an environment with ionising radiation;

4) operations that are in close proximity to uninsulated low/ high-voltage lines or a transformer substation;

5) operations that include the danger of drowning;

6) underground operations such as work in trenches, wells and tunnels;

7) operations in water/underwater or in a caisson requiring an air supply system;

8) operations using explosive gases or liquids (gas tanks, etc.);

9) operations using explosive substances;

10) operations related to lifting, rigging or dismantling heavy construction details (equipment);

11) operations that include the danger of falling from heights;

12) operations that require the checking of labourer health status.

To ensure safety and prevent health risks on the construction site, any employer who has labourers on site must abide by the nations laws and regulations. This requires special attention when working abroad. The employer must ensure proper use of work and protective equipment, ensure that restrictions on the use of materials are followed and obey the orders of the work safety coordinator, if there is one appointed on the site.

6.2   The duties of building contractors

The contractor is obliged to:

❑ follow the requirements and preventive principles of work healthcare and work safety laws and devise a construction site management project during the period of preparation for construction;

❑ prohibit work for labourers who:

   • lack the knowledge and skills of their speciality and the relevant knowledge of healthcare and work safety, and

   • who are intoxicated with either alcohol or narcotics;

❑ inform the technical supervision organisation of a work accident that was caused by a non-conformity to restrictions on the construction or the building as soon as possible;

❑ give any relevant information to the technical supervision organisations representative or any other authorised personnel in order to find the cause of a work accident, in the meantime preserving the scene and outcome of the accident;

❑ enforce systematic internal control of the work environment, during which he or she organises, plans and monitors the company's healthcare and safety situation according to the law or restrictions made by enacted legislation. The internal control of the work environment is an inseparable part of the operations of the company. This control will involve the labourers and will involve the work environment risk assessment. The risk assessment clarifies the work environment's hazards, if need be measures their parameters and assesses the risks to labourers' health and safety, taking into account gender and age discrepancies;

❑ annually review the status of the work environment's internal control and analyse the results. If required, proper adjustments must be made according to any variations discovered. The results of the risk assessment will be documented and preserved for 55 years;

❑ devise a policy (and allocate funds) for the work environment risk assessment in which there are: operations to reduce or avoid health risks, a time schedule and enforcement mechanisms. This policy is to be enforced in every field of activity and at every management level throughout the company;

❑ organise a new risk assessment if working conditions change, if work equipment or technologies are modified or upgraded, if there is new information on a hazard posing a risk to human health, if, because of an accident or a dangerous situation, the risk level has risen or the work healthcare doctor has identified an illness linked to the labourers' work through a health check;

❑ ensure that the labourers working in a danger zone have had special training or special guidance or are being supervised by a labourer who has;

❑ inform an underage labourer, or that person's legal guardian, of the risks and precautions taken to ensure his or her safety;

❑ inform labourers of risks, the results of the risk assessment and the precautions being taken to avoid bodily injury through work environment proxies, members of the work environment council and the labourers' trustees.

❑ implement measures from contracts of employment and collective agreements to avoid physical harm and to neutralise the effects of the risk hazards mentioned earlier. Organise work healthcare and cover the costs;

❑ organise health checks on labourers who might be affected by hazards because of the nature of their work, as defined here or in any other legal act involving the matter, and cover the costs;

❑ appoint labourers fit to give first-aid within the company, bearing in mind the size of the company and its division to sub-units, and organise first-aid training and cover the costs. If the company's sub-units are in different territories or work in shifts, then there must be at least one labourer at all times in the sub-unit or work shift who has first-aid training;

❑ ensure the availability of first-aid kits to every labourer. The first-aid kits must be properly labelled and easily accessible;

❑ transfer a labourer to another field of work or temporarily ease his or her work conditions, according to the laws of employment, if he or she demands it and has a doctor's recommendation;

❑ provide personal protective equipment, work clothing and means of cleaning, if the nature of the work demands it, and organise special training in the use of personal protective equipment;

❑ introduce work healthcare and work safety regulations to the labourers and enforce them;

❑ organise proper training for work a labourer is either starting or being moved to, according to work healthcare and work safety guidelines. Guidance or special training must be repeated if the work equipment and technology is either replaced or upgraded;

❑ devise and authenticate a safety manual for the work being done and the work equipment being used, and inform labourers of how to refrain from polluting the environment;

❑ inform the local Labour Inspectorate of the start, or change, of operations in the field of work of the contractor's company;

Note: The employer has the right to enforce more stringent work healthcare and work safety regulations within the company than are present in the enacted legislation.

6.3   The obligations and rights of the labourer

The labourer is obligated to:

❑ be a part of the creation of a safer working environment, according to the work healthcare and work safety regulations;

❑ follow the work and rest periods announced by the employer;

❑ go through health checks, according to the enforced policy;

❑ use the prescribed personal protective equipment as required and keep them in working order;

❑ ensure that his or her work does not endanger his or her own life or health, or that of a co-worker's, and that he or she does not pollute the environment, according to the employer's instructions and special training;

❑ inform the employer or his or her representative and the work environment proxy immediately of an accident or the threat of one, or a health disorder disrupting work duties or deficiencies in safety protocols;

❑ comply with instructions from the employer, the work environment specialist, the work healthcare doctor, the labour inspector and the work environment proxy, according to the work healthcare and work safety orders;

❑ use the work equipment and dangerous chemicals as directed;

❑ refrain from dismantling, changing or removing safety devices of work equipment or construction without authorisation; use same as required.

The labourer is prohibited from working under the effects of alcohol, narcotics, toxins or psychotropic substances.

The labourer has the right to:

❑ demand proper personal and collective protective equipment from the employer, according to the work healthcare and work safety regulations;

❑ receive information about hazards, the results of the risk assessment, the precautions being taken to avoid bodily injury, the results of health checks and the labour inspectors precepts to the employer in the work environment;

❑ stop working or leave his or her workplace or the danger zone in case of a serious and unavoidable risk of accident; refuse work or stop work that endangers his or her health or that of a co-worker, or does not comply with the requirements of environmental safety, in which case he or she should notify the employer or a representative of the employer and the work environment proxy immediately;

❑ demand a temporary or permanent transfer to another line of work or the easing of working conditions, with a doctor's recommendation;

❑ inform the work environment proxy, the members of the work environment council, labourer's trustee and the construction site's labour inspector if he or she believes that the measures being taken to prevent pollution of the environment are insufficient.

6.4   Ensuring safety on the construction site

6.4.1   General

A safety coordinator must be appointed to the construction site by the main contractor or the owner for the duration of construction. Appointing a coordinator does not relieve the contractor or owner of their responsibilities.

For the duration of construction, the safety coordinator must:

❑ organise and coordinate work safety activities on the construction site;

❑ ensure the introduction of the work safety plan to the labourers working on site and their employers, including subcontractors, sole proprietors, etc.;

❏ check the work safety plan, construction project and adherence to the safety requirements made by technological maps, scheduling them appropriately if there are any amendments to work operations;

❏ make sure that all underground and ground cables, pipes and other installations, including danger zones, are labelled with the proper warning signs, and that the appropriate precautions are being taken;

❏ make sure that the labourers working on the construction site and any other authorised personnel are equipped with appropriate personal protective equipment;

❏ organise regular general inspections on the construction site.

6.4.2 Safety requirements in a work zone

The buildings and workplaces have to have the strength to sustain the work load for the duration of the construction.

Workplaces have to have enough height and square footage to allow labourers to do their work without damaging their health. For every labourer in a workplace, there has to be at least $10\,m^3$ of air space (when calculating air space, the height of a room will be considered to be $3.5\,m$):

❏ See-through walls in close proximity of workplaces and walking routes have to be made of safe materials or protected from shattering and appropriately labelled.

❏ Outside workplaces and walking routes that labourers use must be properly organised so that personnel are not endangered and traffic not disrupted.

- Materials, devices and objects that pose a threat to labourers' health and life must be appropriately and safely stored, and if required, fixed into position.

- Access to spaces built of materials with insufficient strength must be prevented if there are not measures being taken to make the work there safe.

- Labourers must be protected from falling objects, preferably with collective protective equipment. If need be, walkway routes must be covered or access to danger zone prevented.

- Every workplace must have appropriate protective, lifesaving and first-aid equipment in order to prevent or reduce health risks.

- If the workplace has danger zones where there are threats of accident or bodily injury because of the nature of the work, then those zones must be properly labelled and measures must be taken to prevent the access of personnel without special training or guidance.

- The territory, the staircases, the walking routes, and the work and non-work rooms of the workplace must be properly lit. Lights must be placed so that they do not harm the labourers. Lighting must ensure the good visibility of danger signs and emergency shut-down devices.

- The employer must implement measures to prevent or reduce physical health risks from noise, vibration, ionising radiation, etc.

- Labourers doing heavy physical work, working in forced positions for long periods or doing monotonous work have the right to have breaks included in their working time.

❏ Employers must provide suitable working and non-working conditions to ensure safety for underage and disabled labourers by enforcing restrictions according to the enacted legislation.

6.4.3   *Special requirements for assembly works*

Assembly works must be handled in work zones where other work operations are prohibited and unauthorised personnel are forbidden to enter.

During construction, personnel are forbidden from occupying sections on top of which assembling operations are taking place or loads are being moved.

When slinging handleable and installable elements, inventory slings and other cargo capturing devices must be used. These must be made according to an authenticated method, checked and certified. The available slinging manner must prevent the cargo from falling or sliding when lifted and must provide the opportunity of unhooking it from a distance, if the work level from where it is lifted exceeds 2 m.

Swinging or revolving of a lifted construction element must be prevented by binding it with rope.

Openings in ceilings for devices, elevators, staircases, etc., which can be accessed by personnel, must be covered with strong, heavy and immovable shields or be surrounded by railings.

Openings in walls that are bordered by ceilings or work levels/ stages, but also borders of ceilings on top of exterior walls (that are built later) – be equipped with railings.

If the assemblers have to cross from one construction to another, they must have ladders with handrails, overpasses or supports at their disposal.

The assembled element can be unslung only if it has been temporarily or permanently secured in its intended position, according to the project plan.

Assembly is not allowed if the wind is ≥15 m/s, when there is ice, during thunder storms or thick fog or when the visibility across the work place is limited. Vertical panels and other details that have large sail areas must not be lifted if the wind is ≥10 m/s.

6.4.4   Special requirements for work in pits, wells, in tunnels and earthworks and underground

For operations in pits, wells, tunnels and underground, the following precautions must be taken:

❑ The soil has to be properly supported (embankments).

❑ Dangers that may cause workers objects or materials to fall, or that may allow the intrusion of water, must be forestalled.

❑ Every work place must be equipped with a durable ventilation device to provide adequate fresh air.

❑ Labourers must have the means to take refuge safely in case of fire, deluge or fall of materials or collapse of structures.

Before digging operations can commence, the dangers from underground cables and other transmission systems must be identified and brought to a minimum danger level.

Pits, wells and tunnels must have safe exits and entrances.

Piles of soil, materials and vehicles must be kept away from the digging site and, if need be, barriers must be erected around the digging site.

6.4.5   Special requirements for working at height and on roofs

If, while working or moving, there is a threat of falling from a height of >2 m, special safety measures, like railings, safety nets and other such measures, must be used. If using such measures is impossible, because of the nature of the work, then the labourer must be given a safety belt or body harness and be attached to safety cables or ropes. Other methods to ensure worker safety may also be used.

Where the nature of the work poses a serious threat of falling, or the work is being done on top of materials that pose a serious threat if fallen onto, such safety measures must be used even if the height is <2 m.

Railings being used to prevent falling must have a handrail at a height of at least 1 m, a footrail and a rail in the middle at a height of 0.5 m. The rail in the middle can be replaced with appropriate plates or nets. Railings must be placed on the sides of gangways and work stages that have a height of at least 2 m. Scaffolding must have railings if the height of the fall is above 2 m.

If the angle of the roof is <15° and the eaves are higher than 3.5 m, then there must be a barrier with three rails on the edge of the roof. If the work is carried out in good weather conditions and the roof is slip-proof, then the railing must be attached if the edge of the roof is higher than 5 m.

If the angle of the roof is >15° and the eaves are higher than 2 m, then railings or safety nets must be installed, and in the case of a slippery roof, the work area has to be covered with foot supports 30 cm apart.

If the angle of the roof is >35°, then in addition to the afore-mentioned, a railing or a safety net must be installed no further than 5 m from the work area.

If the angle of the roof is >60°, then the railings or safety nets mentioned should not be farther than 2 m from the work area.

If work on the roof is short term and the labourer is using a safety belt or a harness, the stipulations mentioned earlier are unnecessary.

The means of installing and removing safety apparatus onto a roof must themselves be made safe for the labourer.

6.4.6   Special requirements for demolition work

A construction site organisation project must be formed for demolition work that is especially attentive to the work order and the temporary supports of other structures.

The demolition work must be supervised by a qualified person, ensuring that:

❑ before demolition the object being demolished is not connected to any electricity line, nor to gas or water pipes, and that it has no other connections;

❑ when demolishing constructions with asbestos, the standing special requirements are met;

❏ waste and materials liable to cause dust can be lowered from the construction by chute; such loads must be covered during transportation.

Simultaneous demolition work on several floors is forbidden. In addition, it is forbidden to collapse materials on sub-ceilings. Labourers must be protected from falling objects. Areas where such possibilities exist must be defined as danger zones. If need be, covered gangways must be built, or access to the danger zone prohibited.

6.4.7   Ventilation in the workplace

The workplace must be supplied with fresh air. The level of fresh air required is calculated by taking into account the nature of the work, the work methods being used and the physical strain the labourers are under.

Dangerous substances or dust that can damage health, and which is created during the work process, must be removed from the workplace.

The ventilation system being used must be properly maintained and not cause unhealthy drafts.

The ventilation system must be equipped with an automatic control system that notifies personnel in case of malfunction, which could damage labourers' health.

6.4.8   Emergency exits from the workplace

Emergency exits must be clear at all times and allow direct access to a safe zone.

The number and locations of emergency exits is calculated by taking into account the size of the construction site, its location, the work equipment being used and the maximum number of workers on the construction site.

Emergency exits must be properly labelled and equipped with emergency lights to protect labourers coming into danger through a malfunction in the lighting system.

Chapter 7
Requirements for work equipment

Chapter outline

7.1 General requirements

7.2 Mobile work equipment

7.3 Lifting devices

7.4 Dangers from energy

7.5 The usage of work equipment

7.6 Usage of work equipment for temporary work at height

    7.6.1 General

    7.6.2 The usage of scaffolds

    7.6.3 Supports, formwork and heavy prefabricated details

7.7 Work with flammable and explosive materials

The Engineer's Manual of Construction Site Planning, First Edition. Jüri Sutt, Irene Lill
and Olev Müürsepp.
© 2013 John Wiley & Sons, Ltd. Published 2013 by John Wiley & Sons, Ltd.

7.1   General requirements

The employer must ensure that the work equipment – machine, device, installation, transport equipment, tool or any other means – is suitable for its intended purpose and is kept properly and maintained so as to be safe for the duration of its use. If it is not possible to ensure total safety, then measures must be taken to bring the risk to a minimum level.

Using work equipment – working with it, starting it, stopping it, transporting it, moving it, installing it, fixing it, configuring it, cleaning and maintaining it – cannot endanger the user, or anybody else's health, and cannot damage the work and the human environment.

The employer must provide the necessary training and safety guidance to the user before he or she starts using the equipment.

Safety guidance must cover:

❑ information on the dangers, and dangerous situations, that may arise when working with the equipment;

❑ the safety measures to be taken to ensure the safety of the user and other personnel who are authorised to enter the work area;

❑ information on more dangerous work equipment that is in the work area or nearby;

❑ information about changes in the work environment that may increase the dangers coming from the labourers work equipment or the equipment near to the labourer;

❑ instructions on how to handle emergencies.

The employer must prepare and authenticate safety manuals for the equipment being used, according to the equipment manufacturer's manual.

Labourers working with pressure and lifting devices, non-road mobile machinery and other dangerous equipment, must undergo special training organised by the employer, and if required, take periodical refresher courses.

The guidance and training must be repeated when work equipment is changed or upgraded. Each labourer's guidance and training data is registered.

The employer must consult the labourers and the work environment representatives and take into account their proposals to decrease and avoid dangers arising from work equipment.

The employer must ensure that ladders and scaffolds are in working order. Ladders must be checked at least once a month.

The work equipment and its parts – platforms, stairs and other areas used by labourers when operating equipment – must have sufficient strength to withstand the strain of the equipment and must have safety railings; in addition, the equipment cannot cause slipping, stumbling or falling.

The control, guidance and warning means for the equipment must be clearly visible, properly labelled and easy to understand.

To avoid dangerous contact with a moving part of the work equipment, a safety railing or a safety device must be installed to prevent access to dangerous area. Where there is greater danger, the safety railings must be equipped with a means which, by removing the safety railing, stops moving equipment even before the user instructs it to do so.

Any immobile external part of the work equipment that is not guarded but could be dangerous to the labourer must be painted with either alternating yellow and black or alternating red and white stripes.

7.2   Mobile work equipment

Work equipment with an operator (driver or driver and passenger) must have safety adjustments to provide safe driving, including adjustments that prevent accidental contact with the wheels or tread and prevent the driver from falling under or between them.

Forklift trucks must be modified or installed with devices that ensure the safety of the operator if the machine rolls over. Such devices include:

❑ a safety railing or other safety structure over the driver's seat that keeps the driver in place and prevents him or her from falling out, falling under or getting crushed by the machine if it rolls over;

❑ the construction of the forklifts, which must provide enough free space between the ground and the body of the forklift if the machine rolls over.

Any self-moving work equipment that could endanger the labourers with its movements must have:

❑ a start control that prevents unintentional activation of the work equipment;

❑ suitable safety devices that avoid possible collisions of two machines moving on rail-tracks at the same time;

❑ braking and stopping device in case the main control device malfunctions. There must be an easily accessible emergency shutdown device or an automatic system to stop the machine if this is necessary to provide safety.

Mobile work equipment can only be used by labourers with the necessary special training.

The employer must ensure that all the requirements in the manufacturer's manual are met when work equipment is used, serviced and configured. The employer must ensure that before work equipment goes into use, it is correctly assembled and is in working order. The periodical inspection and testing of work equipment is performed according to the manufacturer's instructions or the enacted legislation.

The results of the control and testing of the work equipment are registered and preserved as follows:

❑ The results of the inspection and testing carried out before the equipment is put to use and the results of random inspection and testing are preserved until the equipment is no longer in use.

❑ The results of each periodical inspection and test must be preserved for at least three months after the subsequent periodical inspection or test and the registration of the results.

The results of the inspection and testing of the work equipment must be presented to the national oversight official if he or she so requires.

7.3   Lifting devices

A stationary lifting device must have a sturdy construction and be properly installed, taking into account the force it generates when lifting cargo and the burden it puts on the secured points.

The operation booth of the lifting device must have a clearly visible sign of the nominal load of the device and if required the lifting loads in various positions of the lifting device or various auxiliary means.

Tower crane rail tracks must be grounded from both sides, to prevent possible accidents if labourers are caught in the crane's electric circuit.

When using a mobile or movable lifting device, it must be ensured:

❑ that it is sturdy according to the profile and load bearing capacity of the ground;

❑ that when working near overhead electricity lines, the proper safety requirements are followed.

A labourer cannot be under a load that is being lifted, if it is unnecessary for work operations. Only slingers with special training can take part in lifting operations.

Moving cargo over an unprotected workplace where there are labourers is prohibited. If it is not possible to meet this requirement, other measures must be taken in order to provide safety for the labourers.

Lifting accessories must be selected according to the loads to be handled, gripping points, attachment tackle and the atmospheric conditions. Lifting accessories must be labelled with their technical specifications according to the relevant requirements.

When using a mobile lifting device to lift cargo, the employee must use measures to make sure the device does not tilt, roll over or move by itself from its fixed location, and must ensure that the measures are correctly enforced.

If the operator of the lifting device cannot follow the cargo's whole path visually, there must be a qualified signaller to guide the operator of the lifting device. The signaller must enforce work-organised measures to prevent harm to the labourers from accidental collisions.

If the labourer fixes or releases the cargo by hand, precautions must be taken to ensure this is done safely and that the labourer has a direct or remote control over the lifting device.

If the lifting device is not equipped with a safety device that prevents the cargo from falling in the event of a total or partial power loss, then other measures must be taken against this hazard.

Hanging cargo cannot be unsupervised, except when the cargo is safely secured or access to the danger zone is blocked.

Operation of a lifting device outside must be stopped if atmospheric conditions worsen to a degree where they could endanger the operation of the device or the personnel servicing it.

7.4   Dangers from energy

Electrical devices and instalments on the construction site must comply with enacted legislation.

Electrical installations must be designed, built and used so that there is no risk of explosion or fire. Labourers must be protected from electric shocks and from direct or indirect contact with the source. Protection is ensured by:

❑ isolating, railing or preventing access to current conductive parts;

❑ division or grounding;

❑ discharging or grounding static electricity.

When designing and choosing electrical devices and protective equipment, the relevant properties of every workplace must be taken into account and suitable safety precautions taken, for example regarding the electrical conductivity of workplaces and danger of explosion.

While using work equipment, threats from gas, steam, liquid, compressed air or any other sort of energy must be minimised.

Explosions deriving from substances used or produced by work equipment must be prevented. Prevention of the following is necessary:

❑ concentration of the explosive substances in the air;

❑ combustion of dust and gas mixtures.

The interruption, recovery or variation of work equipment power supply cannot cause a dangerous situation.

7.5 The usage of work equipment

Work equipment can only be used for its intended purpose and in its intended conditions. If work equipment is used in other conditions, the employer must enforce supplementary safety measures.

The position and manner of instalment of work equipment, spaces between the movable and immovable parts, power supply and delivery, and use and removal of peripherals must be safe for both the user and the personnel surrounding the equipment.

If the structure of the work equipment does not allow permanent fixing, and the user, other personnel and their possessions may be unsafe, the work equipment must be firmly secured to a platform using specially designed connections.

If the incorrect assembly of the work equipment parts, gas pipes, steam pipes, liquid pipes or electric circuits could cause a threat, the connection points must be labelled with instructions for assembly and, if need be, the direction in which the part or liquid should move.

During breaks, when any dangerous parts the work equipment may have are stopped, the power must also be turned off.

Equipment operation, control and warning devices must be clearly visible, properly labelled and easy to understand.

Generally, the equipment operating device must be outside the danger zone. Its intentional or unintentional use cannot cause extra danger.

The user must make sure that nobody is in the danger zone before activating work equipment. If that is impossible, the automatic warning device must give out a warning before the equipment is activated. The delay before activation must be enough for workers to leave the danger zone or to use technical aids which minimise danger during the activation or deactivation of the work equipment.

Spontaneous activation, deactivation or change in the work regime must be prevented. These can only happen if the operating device is used. This does not apply to the normal working cycle of an automatic control device.

All work equipment must be equipped with a deactivation device for total and safe deactivation. The deactivation device must be given priority over the activation device so that unintentional activations may be avoided. The operating systems of work equipment must be safe. A malfunction in the operating system or damage to it cannot cause danger. If required, an automatic deactivation device and an electricity cut-off switch must be installed on the equipment.

If the work equipment has a warning to alert users of its dangerous malfunction or break down, the signal given out must be easy to understand and loud, or clearly visible.

7.6   Usage of work equipment for temporary work at height

7.6.1   General

Temporary work at height is taken to be work at over 2 m high when using a scaffold, ladder, rope, hawser or any other temporary work equipment.

Work equipment for temporary work at height must be suited for the work and able to withstand the expected burden. In addition, the work equipment must be positioned so that it allows safe access to the workplace.

Workplaces on top of ladders during temporary work at heights can only be used if the usage of other safer work equipment is not justified because of minimal danger, short period of use or on-the-spot conditions that the labourer cannot change.

Ropes and hawsers can only be used during temporary work at height if the risk assessment shows that it is safe and the usage of safer work equipment is impossible.

Ladders must be positioned so that they remain firm for the duration of use. Ladders must stand on a properly sized, strong and immovable base so that the steps are horizontal.

A hanging ladder, rope ladder excluded, must be attached so that the ladder does not move or swing.

A collapsible ladder must be prevented from slipping by securing the top or bottom of the ladder with equipment that prevents such slipping. An access ladder must be long enough to reach at least 1 m above the accessed level, unless the ladder is stationary.

7.6.2 *The usage of scaffolds*

Scaffolds must be constructed and assembled so that they can be safely installed, used, dismantled, changed and maintained. Generally, scaffolds must be industrial or made by a civil engineer.

All scaffolds must be installed and maintained with their strength in mind, so that they are sturdy for any kind of activity.

If the calculated strength of the chosen scaffold is unobtainable or does not include the relevant construction guidelines, then a strength and stability calculation must be performed, unless the scaffold is installed in the generally recognised standard form.

Scaffolds that are near material or personal traffic routes or cargo lifting zones must be protected from blows, damage and rotation. The danger zone around a scaffold must be isolated with railings and warning signs.

Scaffolds must be equipped with special means to avoid the slipping of supporting parts or other effective solutions. The base must have a sufficient load-bearing capacity and must ensure that the scaffold will stand steadily. Scaffolds with wheels must have measures that prevent random movements.

The size, form and position of the scaffold must be suitable for the specific work operation and be able to carry the load required. It must also provide safety for labourers working and moving on it. The scaffold platforms must be installed so as to ensure that in regular use their constituent parts do not move and there are no dangerous spaces between the vertical railings that prevent falling. The space between a scaffold and a wall cannot be more than 30 cm.

If some of the scaffolds are not ready for use during installation or the scaffolds are being dismantled or modified, they must be labelled with the proper warning signs and access to their danger zones must be prevented.

Scaffolds used in construction must have installation and dismantling plans. Scaffolds can only be installed and dismantled by labourers who have had special training in:

❑ understanding the installation, dismantling and modification plans;

❑ safety measures used during installation, dismantling and modification of scaffolds;

❑ measures used to prevent labourers or objects falling;

❑ safety measures used in bad or worsening weather conditions to prevent damage to the scaffolds;

❑ load-bearing capacity of scaffolds;

❑ other dangers relating to the installation, dismantling and modification of scaffolds.

This special training must be documented.

Metal scaffolds must be grounded so that workers are safe from random electrical current. If the scaffold is positioned on one side of the building, it must be earthed from one place; if it is positioned on two or more sides, then in at least two places.

Scaffolds, ladders and work platforms must be checked before they are put to use on the construction site, including cases where they have been exposed to strong winds, have been under heavy equipment or loads or have been unused for over one month.

7.6.3 Supports, formwork and heavy prefabricated details

Metal and concrete supports and their parts, formwork, assembly details, as well as temporary supports and support walls can only be installed and dismantled under the guidance of a qualified person.

Safety measures must be used to protect workers from the dangers of temporarily unstable structures or structures at risk of failure.

Formworks, temporary supports and support walls must be designed, installed and maintained in such a way that they can bear the load that they are intended to carry.

When checking these structures, special attention should be paid to the support and protection structures.

7.7   Work with flammable and explosive materials

Work on the construction site must be organised so that there is no risk of fire.

Depending on the features of the site's different workplaces, including room sizes and applications, characteristics of substances that are used and stored, the maximum number of labourers, etc., the construction site must be equipped with enough fire extinguishers.

The primary fire extinguishers must be placed in visible and easily accessible places as close to exits as possible, or immediately beside workplaces where fire hazards are most likely to occur.

If there are explosive substances used or stored on the construction site that can release explosive gas or dust when used, safety measures must be enforced to decrease fire and explosion hazards.

There must be instructions on how to act during a fire on the construction site.

Chapter 8
Work healthcare

Chapter outline

8.1   Allowable physical effort

8.2   The usage of personal protective equipment

8.3   Welfare facilities and first-aid

The Engineer's Manual of Construction Site Planning, First Edition. Jüri Sutt, Irene Lill
and Olev Müürsepp.

8.1   Allowable physical effort

Work methods and equipment must be chosen so that they do not overburden the labourer.

When moving weights manually, work healthcare and work safety laws must be followed.

The employer must design and adjust workplaces in which weights are moved manually to be as safe as possible for the labourer. For this, the employer must:

❑ assess risks to the labourers' health taking into account possible risk hazards: the weight of movable loads, their distribution and main measurements, work conditions (characteristics of the working surface – stability, roughness, sufficient space, lighting, body position) and the overall time of the lifting work during a shift;

❑ use safety measures to avoid or decrease any risk that occurs.

The employer must ensure that the moved loads do not exceed the physical capabilities of the employee.

If most of the labourers' work time is consumed with moving operations, the labourer cannot be under 18 years old. Pregnant women and women three months after a pregnancy, and all workers under 16 years old, are not allowed to perform moving operations.

8.2   The usage of personal protective equipment

Protective helmets are mandatory on a construction site. It is recommended that labourers, foremen (brigade leaders) and construction managers (engineers) wear helmets of a different colour to other workers.

On scaffolds, roofs, work platforms and other places where the threat of falling cannot be avoided with other safety measures, safety harnesses with the proper attachment systems must be used.

Generally, protective footwear must be used in construction. During flooring operations or other operations that require kneeling, kneepads must be used.

If work is carried out in the dark or underground, work clothes must have reflectors or reflective strips. For work in places where there is vehicle traffic, labourers must wear a safety vest or safety clothing and if this work is done in the dark, additional reflective strips are required.

When choosing protective equipment, personal protective equipment must be preferred.

8.3 Welfare facilities and first-aid

The construction site should be equipped with enough non-work rooms, for example changing rooms, washrooms, toilets and rest rooms. In case of field work, then, warming rooms and dining rooms and other non-work rooms.

Labourers' non-work rooms must be built and equipped according to the working conditions, number of labourers and gender membership. The necessary non-working rooms are calculated and designed during the construction management project:

❑ Labourers wearing work clothes must have changing rooms and labourers doing field work must have warm rooms and drying rooms for clothes.

❏ Depending on the nature of the work, the labourers must have the opportunity to rest, if required, to ensure the safety and health of the labourers. Rest rooms must be satisfactory in size and equipped with tables and seats with back supports. There is no smoking allowed in the rest rooms.

❏ Depending on the nature of the work, labourer must have the opportunity to use the washroom, which must be equipped with wash basins or showers and hot and cold water.

The labourers must be provided with drinking water, including non-reusable or washable drinking vessels.

The workers must be ensured first-aid from a qualified person if there is an accident or sudden illness on the site.

There must be accessible first-aid kits and eye wash on the construction site. The location of the first-aid kits must be properly signed.

Appendix
Construction site layout symbols

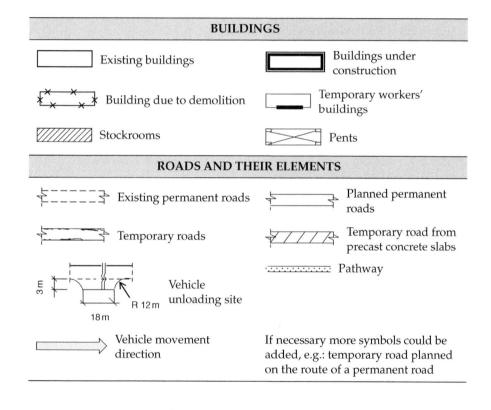

BUILDINGS		
Existing buildings		Buildings under construction
Building due to demolition		Temporary workers' buildings
Stockrooms		Pents

ROADS AND THEIR ELEMENTS		
Existing permanent roads		Planned permanent roads
Temporary roads		Temporary road from precast concrete slabs
Vehicle unloading site		Pathway
Vehicle movement direction		If necessary more symbols could be added, e.g.: temporary road planned on the route of a permanent road

The Engineer's Manual of Construction Site Planning, First Edition. Jüri Sutt, Irene Lill and Olev Müürsepp.
© 2013 John Wiley & Sons, Ltd. Published 2013 by John Wiley & Sons, Ltd.

OTHER COMPONENTS OF CONSTRUCTION SITE LAYOUT

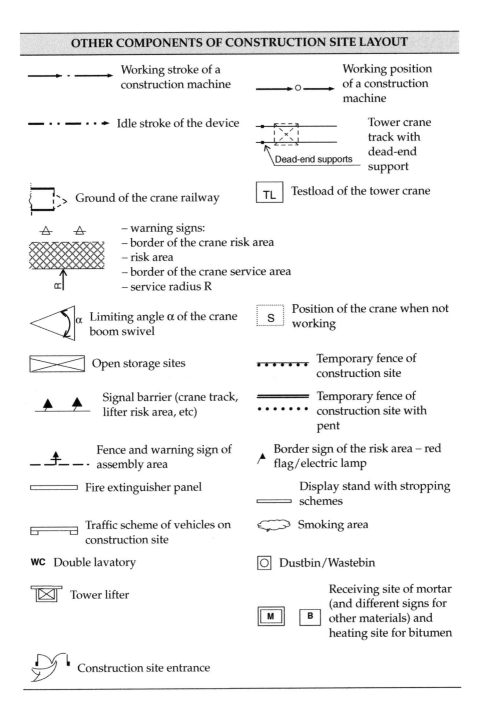

Working stroke of a construction machine

Working position of a construction machine

Idle stroke of the device

Tower crane track with dead-end support

Dead-end supports

Ground of the crane railway

TL Testload of the tower crane

– warning signs:
– border of the crane risk area
– risk area
– border of the crane service area
– service radius R

α Limiting angle α of the crane boom swivel

S Position of the crane when not working

Open storage sites

Temporary fence of construction site

Signal barrier (crane track, lifter risk area, etc)

Temporary fence of construction site with pent

Fence and warning sign of assembly area

Border sign of the risk area – red flag/electric lamp

Fire extinguisher panel

Display stand with stropping schemes

Traffic scheme of vehicles on construction site

Smoking area

WC Double lavatory

Dustbin/Wastebin

Tower lifter

M   B Receiving site of mortar (and different signs for other materials) and heating site for bitumen

Construction site entrance

WATER SUPPLY AND SEWERAGE

——**W**—— Existing water conduit | ——**TW**—— Temporary water conduit
—— **S** —— Existing sewerage | —— **TS** —— Temporary sewerage
——◑—— Fire hydrant well |

ELECTRICITY SUPPLY

——TE –○—— Temporary electric aerial line on poles | —— TC—— Temporary underground cable line
—TEP–○— Temporary cable line on poles | —‖————‖— Cable line on trestles
◄——○——► Existing aerial line with a voltage of ≥10 kV | ◄——○——► Existing aerial line with a voltage of ≤10 kV
—————○————— Existing aerial line of street lighting | ▬▬▬ Input-distribution switchboard
▰▬▰ Distribution switchboard for switching power and lighting equipment | ▭▮▭ Power supply of the crane
▲ TT N° Temporary transformer substation | ══════ Cable in tube
• N a/δ Floodlight pole, where N – number on the layout, a – output, δ – installing height |
⚡ Floodlight or floodlight type lighting | ⊗ General-purpose lighting
Existing trees for preserving | Existing trees for taking down

TRAFFIC SIGNS

⊖ **No entrance** | ⑩ **Maximum speed**

RISK SIGNS

△ Warning sign with an explanatory text, eg 'Crane in operation!', 'Falling objects!', etc

△ Other risks!

▫ Obliging signs. e.g. 'Work with safety belt!'

SUPPLEMENTARY SYMBOLS

Up–coming building	E Entrance
Temporary fence that coincides with the planned permanent fence	Board with construction passport
─G─ · Existing gas piping	Junction of temporary utility network
Temporary light pole	Permanent/planned light pole
Floodlight on pole	Floodlight on transportable tripod

Bibliography

Bauer, H. (1994) *Baubetrieb 2: Bauablauf, Kosten, Störungen*. Aufl. Springer-Verlag, Berlin.

Construction Site Workplace Safety Plan. Health and Safety Risk Management. www.safety.com.au (accessed on 5 February 2013).

Ferguson, I. & Mitchell, E. (1986) *Quality on site*. B.T. Batsford Ltd., London.

Health and Site Executive. http://www.hse.gov.uk/construction/areyou/cdmcoordinator.htm (accessed on 5 February 2013).

Hedfeld, K.-P. (1992) *Wie organisiere ich meinen Baubetrieb Richtig?* RKW RG-BAU, Eschborn.

Illingworth, J.R. (1994) *Construction Methods and Planning*. E & PN Spon/Chapman& Hall, London.

Mantscheff, J. (1991) *Bauvertrags- und Verdingungwesen. VOB Teil A u.B*. Werner-Verlag, Düsseldorf.

Peurifoy, R.L., Schexsnayder C.J., Shapira A. & Schmitt R. *Construction Planning, Equipments and Methods*. McGraw-Hill Education, New York.

Temporary Work Design. http://www.twd.nl/contact.html (accessed on 5 February 2013).

The Management of Temporary Works in the Construction Industry. http://www.hse.gov.uk/contact/index.htm (accessed on 5 February 2013).

Dikman, L. Organizacija, planirovanije i upravlenije. Moscow: Vyshaja Shkola, 1982.

The Engineer's Manual of Construction Site Planning, First Edition. Jüri Sutt, Irene Lill and Olev Müürsepp.
© 2013 John Wiley & Sons, Ltd. Published 2013 by John Wiley & Sons, Ltd.

Index

allowable physical effort, 170
assembly work safety/assembly works, 68–76, 149

bidding stage, 15, 16, 20, 22, 23, 25, 26, 29, 46, 50
bill of activities, 5, 7, 8
bill of quantities, 2, 5, 7, 8, 17, 18, 35, 38, 46

calculation of duration, 8, 39, 41
construction site lightings, 24, 49, 126–7
construction market, description, xi
contractor, xiii–xv, 13, 17, 23, 41, 43, 47, 123, 138, 139, 141, 144, 146
    responsibility, 138, 141, 146
    technological possibilities, 17
cost estimation, xiii, 1, 15, 23, 25, 26, 29, 41, 46
crane danger and impact areas, 64, 65, 68, 69
crane track, 55, 59–62, 64, 65, 69, 72–4, 174

danger area (zone), 11, 19, 32, 33, 52, 53, 65, 66, 68–71, 89–96, 98, 142, 146–8, 153, 161, 163, 164, 166

demolition works / demolition work, 18, 21, 64, 137, 152, 153
design documents, 6–7, 12, 20
design phase, xiii, xv, 6

explosive materials, 140, 155, 162, 168

fencing, 11, 13, 20, 23, 24, 26, 32, 42, 45, 47, 64, 65, 100, 135, 136
first aid, 114, 143, 148, 171–2
flammable and explosive materials, 168

geometrical parameters on site plan, 34, 54, 59–63, 71

heating and power supply, 48, 116–25

impact area, 51, 64, 68, 69, 77, 87
impact of power line, 52, 91, 92, 94
initial data, xiii, 5, 28, 29, 46, 122, 124

labourer, 138–54, 156–62, 165–8, 170–172
labourer's responsibility, 138, 144
lifting devices, (equipment), xiii, 10, 26, 31, 32, 47, 55, 64, 93, 157, 160

The Engineer's Manual of Construction Site Planning, First Edition. Jüri Sutt, Irene Lill and Olev Müürsepp.
© 2013 John Wiley & Sons, Ltd. Published 2013 by John Wiley & Sons, Ltd.

lifting parameters of crane, 53, 55, 56, 72, 81, 84

lighting, xiii, 24, 26, 32, 33, 40, 45, 49, 100, 122, 124–7, 170, 175

load take up device, 47, 53, 55, 56, 75, 82, 100, 130, 131, 134, 135

methods of calculations, 124

mobile crane, 42, 47, 51, 77–9, 81, 83–91, 94

mobile equipment, 158–9

network chart, 29, 30, 36, 37, 39–41, 43

owner's responsibility, 138, 139, 146

personal protective equipment, 143, 144, 147, 170–171

positioning of cranes, 51–5, 57, 60, 64, 74, 75, 85, 88, 94

power supply, 26, 33, 48, 100, 121, 162, 163, 175

productivity, 27, 38, 40, 43–5, 130

resource allocation, xiii, 3, 29, 33, 41

restrictions, 8–12, 38, 52, 53, 88, 96, 97, 105, 122, 141, 149

safety of underground works, 140, 147, 150, 171

safety requirements on site, 137

scaffolds, 155, 157, 165–7, 171

scale of layout, 20, 30

sequence of procedures, 32, 35

shift, 17, 23, 30, 33, 38, 40, 41, 47, 74, 130, 143

simultaneous operations, 51, 52, 71, 73–7

site inspection, 5, 8, 9, 14

site layout, 1, 13, 15, 16, 19–21, 25, 28–35, 41, 46, 53, 101, 107, 108, 111, 114–16, 122, 173, 174

site lighting, xiii, 24, 26, 32, 33, 40, 45, 49, 100, 122, 124–7, 170, 175

site storage, 19, 96, 99, 105, 108

specifications, 5, 7, 8, 161

technological model, 29, 36, 39, 41

temporary building, xiii, 1, 2, 10, 19, 21, 24, 26, 31, 33, 37, 42, 44, 48, 99, 106, 111, 113–15

temporary facilities, xiv, 19, 31

temporary heating, 99, 116–20

temporary power supply, 26, 48, 100, 121

temporary road, 2, 9, 11, 19, 23, 25, 26, 31, 47, 72, 99–104, 173

temporary water supply, 18, 23, 26, 48, 99, 115

temporary works, xiv, xv, 2, 15, 17, 23, 25, 27, 46, 50, 177

    cost classification, 2, 17, 50

    estimation, 23–7

tender, 6, 35

time schedule, xv, 1, 7, 13, 15–17, 21–4, 29, 33, 35, 41, 46, 122, 129, 142

tower cranes, 22, 31, 32, 42, 47, 51, 53–7, 59, 62, 63, 67–9, 71, 72, 74–7, 83, 85, 87–91, 96, 97, 105, 160, 174

transport on site, 100, 127–30

underground work, 140, 147, 150, 171

utility network, 2, 6, 11, 19, 23, 24, 30, 31, 42, 100, 101, 113, 114, 176

water supply, 18, 22, 23, 26, 33, 37, 44, 45, 48, 99, 115, 175

welfare facilities, 169, 171

work at height, 137, 140, 151, 155, 164, 165

work classification, 43, 44, 50

work equipment, 155–9, 162–5, 170

Printed and bound by CPI Group (UK) Ltd, Croydon, CR0 4YY

27/10/2024

14580213-0002